I0760174

“The world's astronomers, under the auspices of the International Astronomical Union (IAU), have concluded two years of work defining the difference between ‘planets’ and the smaller ‘solar system bodies’ such as comets and asteroids. If the definition is approved by the astronomers gathered 14-25 August 2006 at the IAU General Assembly in Prague, our Solar System will include 12 planets, with more to come”

-- https://www.iau.org/news/pressreleases/detail/iau0601/

Twelve planets! This was the resolution that was intended to be approved at the IAU general assembly. It was an exciting and happy time. Humanity was gaining new territory to spark our imaginations. As you can see, this wasn’t a haphazard event. It took two years of effort. A lot of work had gone into it. The vote of approval was only a formality to make it official. So what went wrong?

Never underestimate the ability of a small group of committed individuals to change the world.

— Margaret Mead

And that’s what happened, but not for the betterment of humanity. A small group of astronomers fought a petty power struggle over what they thought should be important in our Solar System, and what should not. They seized the event in Prague and forced their opinions on the rest of us.

The vast majority of planetary scientists disagreed with them. Most were either unable to attend the vote or didn’t realize it would be such a pivotal event. Afterward, they let their thoughts be known, but it was too late to change the outcome of the vote. That’s not how real science is done.

This book only briefly discusses the internal politics of the IAU. It is more about the science of astronomy, the interplay of objects in the Solar System, and the words and definitions we use to describe it all.

It includes an honest categorical analysis of the largest Solar System bodies and a lot of the smaller ones too. It explores how scientific definitions work, other classification schemes, and how we use technical jargon alongside common everyday language.

The conclusion is simple; it is legitimate and accurate to call Pluto a planet. If you love science, astronomy, and space exploration, you should be calling Pluto a planet too. The proof is in the pages ahead.

Welcome Back

Pluto!

How the IAU got it wrong, why they did it, and how we can help them get it right.

Ron L. Toms

RLT Publishing
www.RLT.com

Published by RLT Industries.
www.RLT.com

Hardback ISBN: 978-1946767066
Paperback ISBN: 978-1946767059

Cover Image Credit: NASA/JHUAPL/SwRI

A Planet By Any Other Name

Pluto, 'tis but thy class that is my enemy;
Thou art thyself, though not a planet?
What is a planet? Is it not size, nor mass,
nor geology, nor density, nor any part
belonging to a celestial object?
O, the inadequacies of this taxonomy!
For that which we call dwarf,
by any other name, would be as sweet a place
to visit, to map, to investigate, and perchance, inhabit.
So Pluto would, were it not dwarf called, retain
that dear quality which it owes without such elfin classification.
Pluto, doff thy dwarfness, and for that diminishing which is no part of thee
Take our imaginations that we might yearn to explore thine world.

Contents:

Foreword

As a lover of science, with nothing but respect for the scientists and researchers who do all the heavy lifting that I get to read about, I had originally accepted the IAU's 2006 resolution regarding the status of Pluto. After all, they were the experts. Or so I thought.

At the time I also had kids in elementary school. I did my best to fill them with a sense of wonder about the greater cosmos and the mysteries of the unknown. I saw first-hand how the reduced scope of our planetary system diminished their interest in it. This made me sad, but science doesn't care about our feelings. I let it go.

Meanwhile, new species of animals were being discovered and reported in the news. Species thought to be extinct had been sighted. My youngest daughter shifted her scientific interest from astronomy to biology, and I didn't think anything more about it.

Then, in 2015, we saw close-up images of Pluto for the first time. The debate over its planethood was rekindled. I was fascinated with the images. I was also disappointed that my kids showed no interest. "It's not a planet," they said, as if that explained everything.

I tried to let it rest, but something about it was smoldering in my subconscious. Occasionally arguments would flare-up in the on-line forums I follow. I got involved.

Using pseudonyms, I poked and prodded, testing the logic on both sides. The issue wouldn't leave me alone. My subconscious woke me up at night with fresh insights and questions. Could the experts be wrong? I felt like I was missing something. Exactly what it was, I didn't know. But I needed to find out.

"If you want to understand something, write a book about it."
— Yogi Bhajan.

So here we are, and I've got news for you—

To summarize this whole book in a single paragraph: It is perfectly legitimate, *and proper,* to call Pluto and Ceres (and Eris) planets. Why?

Because that's exactly how the true experts in the field use the word. It is what the vast majority of planetary scientists confidently and openly call them. The famously flawed definition was the result of a small minority of astronomers who leveraged a procedural protocol to force their opinions onto the world.

That's not how science works, and it's not how definitions work. If you want all the specific details to back that up, and find out how it all happened and how to resolve the problem, you'll have to read the rest of the book.

Why is this important?

We are in the midst of a revolution just as monumental as the shift from the Geocentric to the Heliocentric model of the Solar System. But the words we use to describe its various parts haven't changed. They are no longer adequate. This needs to be fixed.

Whether Pluto is a planet or not depends entirely on what characteristic you're using the word *planet* to represent. And like most commonly used words, there are many competing ideas, some more useful than others.

This book was written in 2021. One hundred years after the 1,000th asteroid was discovered and Edwin Hubble proved that other galaxies exist. The past 100 years of astronomy have fundamentally shifted our understanding of the cosmos, and more importantly, our perspective.

The ancients thought our singular solar system was the entire universe. Until 1921, most people thought our singular galaxy was the entire universe. Today we know there are *billions* of galaxies in the cosmos. We know about black holes, neutron stars, exoplanets, and so much more.

We know there are *millions* of asteroids and comets and other things. within our own solar system. And we are no longer limited to looking at them as mysterious points of lights in the sky. Now we have the capacity to actually go there, to measure and test and sample them, and communicate the results back to us here. This has changed everything.

We'll explore all the reasons why a certain sub-set of scientists are so adamant that Pluto cannot be a planet, including reasons that are not contained in the definition. We'll also hear the other side from equally respected scientists who disagree.

Nature does not care about the nice tidy categories we impose upon it.

— Stefan Milo

How can we know the difference between a minor planet and a major planet? This is a lot like the question of what is a color, and where is the boundary between yellow and green.

Imagine a committee is formed and a vote is taken. They decide there's no such thing as Yellow (for good reason, which we'll get into), and the name Yellow should be removed from the list of colors. This is actually a good analogy for what happened to Pluto.

In nature, there are more shades of color than we have names for, just as in space there are more objects floating around than we have names for. There are invisible colors—wavelengths of light beyond our ability to see, high and low energy colors, different kinds of transparency and polarization and other qualities we may have yet to discover. Which of these properties should we take into consideration when naming the color of a thing?

Most people can only name the six so-called primary and secondary colors, plus black and white, plus a handful of shades of those. The English language has eleven common color names; red, blue, green, yellow, orange, purple, violet, brown, black, pink and grey, but Wikipedia lists thousands. You might guess what color 'greige' is, but what about 'fuzzy wuzzy'? How about 'shiraz'? (And what does this have to do with Pluto? We'll get to that too.)

A trained human eye can distinguish a million different shades of color. Scientific instruments can find billions. Imagine the confusion if each one had its own name. How should we define the bounds of what is, or is not, a color and what is merely a shade of a color?

Astronomers use the images from telescopes to find the dots in the sky and differentiate them from each other. But this is not a complete picture of what's actually out there. It's like looking at images on a computer screen for yellow.

Have you ever seen anything yellow on the Internet? If you said "yes", I'm afraid you are wrong. What you thought was yellow is actually an array of red and green pixels that give the illusion of yellow from a distance. But there is no yellow there. Use a magnifying glass and you'll only see red and green dots on your computer screen. There is no yellow anywhere on the Internet regardless of what our perceptions tell us. Technically and scientifically, it's reddish-green.

What are we going to do with that information? Should a small group of scientists publicly declare that yellow is no longer a color? Can we honestly call yellow a color? The question is tricky because the answer depends on how we use the word 'color' and in what context.

Are we talking about a perception, or something we can reduce to fundamental, measurable quantities? If the latter, then there is no yellow. If the former, there is. So which do we choose? We can, in fact, use both. But we must be mindful that these are different things.

Languages—specifically, the words we use to identify and name things—are imprecise, prone to drift in meaning, but also are the major force that shapes our perception of the world. We think, daydream, communicate with each other, choose careers, vote and connect with the rest of humanity with sloppy, imprecise, perception-based language. Words and definitions are nothing more than symbols representing ideas, and ideas are prone to change.

Colorless Green Ideas Sleep Furiously
— Noam Chomsky

But science is more critical and exacting. It must be. Where exactness and an unwavering persistence of meaning are required, we use a system of taxonomy, often augmented with numbers. So, red on the internet is RGB-255,0,0 but few people ever say things like, "I want a 255-0-0, car." (If they did, most people would assume they're referring to the engine). Green is RGB-0,255,0 and 'Yellow' is denoted as RGB-255,255,0.

Since there's more to nature than the internet, is it valid to restrict our definition to only the images on our computer screens?

In the case of the definition of the word planet, that's exactly what the IAU did. Their definition only applies to our single specific solar system as seen through a telescope's view-screen. It expressly *does not apply* to any other planets in any other solar system. This is not a valid scientific definition for a fundamental concept.

In the case of yellow, there are other color systems unconstrained by what a computer monitor can display, such as CMYK. Notice that the 'Y' in

CMYK stands for yellow, which apparently *does* exist in pigment form. Two different systems exist for color, one additive, one subtractive. They can both work side-by-side without any discord or dissent as long as no one tries to claim authority for the entire definition of the word.

Can we resolve this issue once and for all? Ultimately, all the ambiguity over 'what is a color' can be resolved by stepping outside the system and looking at *how* we look at it. Pure, raw color exists as a broad range of electromagnetic (EM) waves. The human eye can detect three short ranges of these wavelengths via three different types of sensors on our retina. These sensors convert the intensity of the EM wavelengths they are attuned to into a signal that is sent to the brain. We subconsciously interpret various combinations of those signals as single, unique colors. (There are some people who have four color receptors, known as tetra-chromic vision. But they are rare. Some species of animals can detect dozens of colors. Human vision is pretty lame compared to the eyes of squid, eagles and shrimp.)

In nature, there is a wavelength of EM radiation that corresponds with the name yellow. But our technology cannot recreate that wavelength on a computer screen. Only our perceptions can.

Here's the point— If we can legitimately call things yellow on the internet, then our perceptions matter.

Similarly, what makes something a planet? Ancient astronomers were limited by their perspective. They only knew the planets as points of light in the nighttime sky. Most importantly, these points of light moved in relation to the other, non-moving points of light. A planet was defined not by what it was made of, but by the simple fact that it was able to move.

Today, we know there are billions, maybe trillions of things in motion around the Sun that cannot be seen from Earth. It's a vast spectrum of sizes, masses, composition, temperatures, densities and other properties unimaginable to the ancient astronomers who set the precedents for the words we use today. Many of these things have even been measured and sampled by probes. Our technology has given us a new perspectives, and our understanding has changed dramatically. We are no longer limited to thinking of them as things that move in the sky. They are more than just bright points of light in an orbit, they have real physical qualities that are unique and independent from the paths they take. We have metaphorically stepped away from the computer screen and opened the doors to nature.

Except, we wouldn't get an up-close view of Pluto until nine years *after* the vote for that ill-conceived definition took place. The vote was based on incomplete knowledge—a gap that would soon be filled—and *everyone knew it.*

In a profession where discoverers are allowed ten years just to pick a name for a newly discovered object, what was the rush? Was there a significance to the timing of that vote?

In fact, there was. The vote was originally nothing more than a procedural formality intended to confirm Eris as a new planet, re-establish Ceres as a planet, and retain Pluto as a planet. The wording of the original press release made this clear:

"The world's astronomers, under the auspices of the International Astronomical Union (IAU), have concluded two years of work defining the difference between 'planets' and the smaller 'solar system bodies' such as comets and asteroids. If the definition is approved by the astronomers gathered 14-25 August 2006 at the IAU General Assembly in Prague, our Solar System will include 12 planets, with more to come"
-- https://www.iau.org/news/pressreleases/detail/iau0601/

Few people stuck around for the final hours of that week-long meeting in Prague. Most of them were anxious to get home to their families. This vote was a simple formality that wasn't supposed to be a big deal. But a small handful of astronomers took advantage of the situation to force their personal perspectives on the world.

But words and their definitions are, first and foremost, a part of our common language. We cannot ignore this fact. So I've decided to start this book by exploring some interesting words and their meanings. They are often not as precise as we like to think.

In later chapters we'll deconstruct the IAU's official definition (with the help of thoughts and words from actual astronomers and astrophysicists), followed by an analysis of different things in the solar system, different types of orbits, and several planetary scenarios from other solar systems.

Finally, I've included some thoughts about how we classify things in the Solar System, and what we're trying to accomplish with those classifications. The question of whether or not something is a 'planet' is tricky because the answer depends on who is using the word, in what context, and for what purpose. Does the word 'planet' belong to everyday language, or is it part of a precise taxonomy constructed to improve the scientific methods of astronomy, planetology and astrophysics? What about the other scientists and space probe engineers who are not specifically astronomers? Their opinions are relevant too.

Let's explore that furiously sleeping colorless green idea-space…

Section One

Words and Definitions

Chapter One:
Houston, We Have a Problem.

Before we begin, we must acknowledge that it is trivially easy to be a critic. It is tremendously harder to create something original, and harder still to create something that's both original and useful.

The IAU saw a real and legitimate problem and tried to solve it with an original solution. They have done the hardest part. We must respect that and give them credit for the attempt.

But what they did wasn't exactly useful. In the truest spirit of the scientific method, we're going to challenge their work. We're going to explore the flaws in their definition of the word planet, then we'll explore some of the many ideas floating around for improvement. If we fail to accomplish anything with all that, at least we tried. That's all any scientist can do.

And now, on with the show.

I'm sure you're well aware, in August of 2006 the International Astronomer's Union (IAU) issued 'Resolution B5'. It was, for the first time ever, a formal scientific definition for the word Planet in the Solar System.
— https://www.iau.org/static/resolutions/Resolution_GA26-5-6.pdf

The result of that vote was simple and singular. Pluto was excommunicated from the family of planets. Their newly minted definition served no other purpose than that. To be fair, the IAU had a logical reason for what they did. Every profession has its own industry-specific nomenclature used for their own industry-specific operations. It is a necessary function of their jobs. But in this case, this particular definition carried a lot of dissent within the ranks of the IAU, also among the many planetary scientists who are not astronomers, and also among the public who are generally knowledgable about thc subject.

And still, to this day, the debate continues. This book is an attempt to clear up some of the confusion. I know, I know… Don Quixote here, tilting at windmills. But bear with me.

As Neil deGrasse Tyson said,

> **The stated authority of a thousand is not worth the humble reasoning of a single individual**
>
> — Neal DeGrasse Tyson, (pg 127, The Pluto Files)

We'll get to the humble reasonings of actual experts in later chapters, but first we'll have a little fun with words. This is a slightly silly exercise, but it's important. It illustrates that we use words differently from their 'official' meaning all the time, and there's nothing wrong with that.

In fact, many members of the IAU still boldly and confidently refer to Pluto as a planet, even in their scientific papers written for publication, all without suffering any adverse consequences at all.

The controversy over whether Pluto is a planet has clouded one simple fact; no industry or organization owns or controls the usage or meaning of any word in the public vernacular. Not lawyers, not engineers, not biologists, not textile workers, not the military, and not the IAU, as we will see in the next few chapters.

Even the professional lexicographers who compile and edit dictionaries will tell you, not even they have the power to decree what a word shall mean. The common usage of a word reigns supreme. You and I and everyone we know determine its meaning via our communications with each other, and no one else can supersede that.

This does not prevent the IAU from recommending their own industry-specific use of the word. But as we'll see, common language and industry-specific terminology are rarely aligned—even within the sciences. The IAU can claim to have ultimate authority over the word *planet*, and expect us all to fall in line, but the venerable Neil deGrasse Tyson has something to say about that too.

> **If the world is something you accept rather than interpret, then you're susceptible to the influence of charismatic idiots.**
>
> — Neil deGrasse Tyson, *The Pluto Files.*

That may have been an unfortunate choice of words on Mr. Tyson's part. I want to be clear from the start that I have the utmost respect and admiration for Mr. Tyson and the astrophysicists of the world. It is not my intent to tell them how to conduct their business. I only wish to participate in the discussion around the word *planet* on behalf of the rest of the English speaking world. I claim this right as a verified speaker of English, also an occasional writer of books in said language, and therefore have a vested interest in the meanings of its words, as do each and every one of my readers and anyone else with an interest in the subject.

I am also an admirer, promoter and defender of the science of astronomy, as demonstrated by my entrepreneurial background. Briefly, I designed and constructed, and for three years operated one of the most unique planet-walks in the world. It was an exact scale model of the inner Solar System built upon open, undeveloped land, free of visual obstacles or distractions. It was a place where kids could visit and explore the space between planets while walking *at the speed of light* and see the true size and distance between all the inner planets from any vantage point in the park.

The park is no longer there, but the old web site can still be seen at

http://www.SpaceTimePark.com

And so, I have taken it upon myself to investigate the history, logic and bigger picture of the problem around what is, or is not, a 'planet.' This book is nothing more than the humble reasoning of one individual, intended as nothing more than meaningful food for thought to anyone adventurous enough to sample it.

Chapter Two:
A Rose By Any Other Name

And now, submitted for your amusement, a play of words in one act:

What will happen when an exasperated teacher and an unusually bright student with an attitude butt heads over the issue of planets? Our story takes place at an ordinary looking parent-teacher conference, with the exception that this meeting is being held somewhere deep within

The Cosmology Zone…

Prologue:

Young Billy was the kind of kid every parent would be both proud of, and also perpetually embarrassed by. He was the perfect smart-ass—the kind who was not afraid to speak his mind and who was also usually right. So his mother and father were not surprised when Billy's 8th grade teacher called them in for a parent-teacher meeting.

They entered the classroom where Billy and his teacher sat at their desks, silently facing each other. The kid had a look on his face that made his mother anxious, and his father curious. It was clear to them he had something devious on his mind.

Act one:

"Hello Mrs. Pedante, what's the problem today?" Billy's mother asked the teacher as she and her husband took their seats.

The teacher let out a sigh, "Billy won't stop telling the other kids Pluto is a planet," she explained, "Pluto is not a planet. He's confusing them and it needs to stop."

An awkward silence filled the room while Billy's parents traded glances. His father recognized his son's body language from when they played chess together. The kid was already three steps ahead, waiting for everyone else to catch-up. All his father needed to do was open the gate.

"Billy," he said. "Why do you think Pluto is a planet?"

The boy relaxed in his chair, retrieved his lunch bag from his backpack and opened it. He took out a sandwich and a bag of strawberries. "Because it *is* a planet," he replied.

"No." Mrs. Pedante blurted. "It's a dwarf planet."

"A dwarf, *what*?" Billy asked.

"Now don't start that again!" she said.

"It's a dwarf *planet*." Billy asserted. He took an unusually large strawberry from the bag, bit it, then studied the hollow of its interior.

Billy's dad counted the seconds in his head. *Four, three, two...*

"Do you drink coffee, Mrs. Pedante?" Billy asked as he chewed the strawberry. He set the half-strawberry aside and opened the sandwich to peek between the slices of bread.

"What's that got to do with anything?" she responded.

Without looking up he said, "Coffee beans. They're not beans you know. It's actually the seed of the coffee cherry, which isn't actually a cherry. Are you going to make us all say coffee-fruit seeds now too?"

He closed the sandwich and took a big bite out of one corner.

Billy's mother spoke up, "Now Billy, that's different. This is science."

"I know!" Billy quipped, his voice muffled by a mouth full of sandwich, "Like Biology. Biology is a science. Like this strawberry that isn't a berry. It's not even a fruit." He held up the half-strawberry, aiming the bitten end toward the teacher. He waited for Ms. Pedante to react. She did not disappoint him.

"Don't be ridiculous," Mrs. Pedante said, "Of course strawberries are a fruit. You see what I have to deal with?" she implored.

A smile crept over Billy's father's face. *Here it comes.*

Billy swallowed hard and took a breath, "There's no seeds inside. The seeds are on the wrong side of the skin. Technically, strawberries aren't a fruit. They're not berries either. And a pineapple isn't a pine and isn't an apple, but it actually *is* a berry. But no one calls it tropical-fruit berries. Coconut milk isn't milk." He swallowed again, then peeled back the top of his sandwich for everyone to see.

He spoke to the sandwich, "The peanut butter in my sandwich isn't butter, and peanuts aren't nuts. Neither are walnuts, by the way. Wall-*nuts*. Does that bother anyone?" He finally looked at the teacher and said, "If we can say those things then why can't we say Pluto is a nice little planet where I can land my spaceship and go exploring?"

Billy closed the sandwich. He only waited long enough to let the point sink in, but not long enough for anyone to respond. He continued, "My parents drove here using gas that isn't a gas. My grandfather, who wasn't my grandfather, ate sandwiches made with head-cheese, which isn't cheese and is kinda gross, but that's just how grandfathers are. Panda bears are not bears. And tomatoes are not a vegetable, but ketchup that's made from tomatoes *is* a vegetable."

Mrs. Pedante's jaw dropped. She quickly recovered and stated matter-of-factly, "Ketchup is *not* a vegetable."

This time Billy's mother came to his defense. "Actually, according to the US Dept. of Agriculture, tomato paste is technically classified as a vegetable. We were talking about this just last week, how sometimes things get defined in funny ways for technical purposes, but that it doesn't mean we all have to use the word that way."

"And I'm adopted," Billy's father added with a smile.

Ms Pedante sneered, "I suppose you think a dolphin is a fish."

"No, of course not," Billy replied. He took another bite of sandwich and talked through the dryness of the bread and peanut butter, "That would be stupid." He considered his position for a moment, chewed, then added, "A dolphin is clearly a mammal."

Billy's dad decided to test his son's logic, or throw him an easy pitch for a home run. He wasn't sure what the kid had in mind. "Why can't we call a dolphin a fish? It lives in the ocean, doesn't it?"

Billy knitted his brow as he thought and began slowly, still struggling with the stickiness of the peanut butter in his mouth, "Because… it's… it's not about *where* the thing is. It's about *what* it is. The only reason they say Pluto isn't a planet is because of *where* it is, at the edge of the Kuiper belt, not because of *what* it is."

"Pluto isn't a planet because there's other stuff in its orbit." Mrs. Pedante explained.

"That's like saying saltwater isn't water because it has salt in it. You can't drink it, so it must not be water." He swallowed hard again. "No one cares if I say strawberries are the best fruit or that can of mixed nuts is mostly peanuts or mom likes those dark roasted coffee beans. But If I say my favorite planet is Pluto, everyone makes a fuss."

“Well,” Mrs. Pedante said, “I’m afraid the astronomer’s union has more authority on the subject than you do.”

“What about the dictionary?” Billy asked as he reached for his thermos of water, “Isn’t that a higher authority? The dictionary says Pluto is a planet.”

Out of the park, Billy’s father concluded. *Checkmate.*

— End scene. Close Curtain. —

Epilogue:

Suffice it to say, there was no resolution that day. Billy’s answer to the question of Pluto’s status was marked wrong on the test, because his teacher has the authority to do that. It wasn’t enough to change his letter grade, but the bad mark got under his skin anyway. He vowed to become a famous astrophysicist one day and set the record straight.

Chapter Three: What Are Words For?

Here's a short list of a few common words that don't mean what most people think they mean, but that's probably not going to change how anyone uses them.

Argument. You might think two people yelling at each other are having an argument. But a boisterous disagreement is not an argument. According to the lawyers and philosophers of the world, an argument is a well-considered explanation. This book is an argument.

Guns. Anyone who spent time in the U. S. Navy or the United States Marine Corps knows that a hunting rifle (or worse, a pistol) is definitely *not* a gun. In boot camp you will get in trouble for saying it is, even accidentally. A true gun has to be mounted on a ship or some kind of vehicle. It's what lay people would call heavy artillery. The thing you can carry is a personal weapon, not a gun.

Plastic. What we call "plastic" is actually a polymer. To a materials scientist, plastic isn't a thing, it's a property of a thing. Metals, glass and even rocks can become plastic when they get hot enough or are subjected to the right kinds of pressure. Polymers are commonly called "plastic" because of their ability to be formed and molded into any shape.

Thread. You might think the fibers in your jeans and t-shirt are threads, but don't dare say that to a textile worker! The proper term is yarn. That's right—t-shirts, blue jeans, suits, socks… They're all made from *yarn.* Thread is strictly a verb. It's what you do with the yarn. You thread the yarn into the knitter.

Organic. Did you think this word meant "all natural," maybe indicating an absence of chemicals like preservatives, pesticides or herbicides? Sorry, organic chemistry is anything and everything that involves carbon compounds. This includes living and non-living carbon-based chemistry, such as petroleum and things derived from petroleum, like preservatives, pesticides and herbicides. They are all organic.

The science of words is called lexicology. The social sciences are sociology. The study of insects is entomology. Life science is biology. Weather science is meteorology. The science of rocks and their formations is geology. But if you call the study of stars and planets *astrology,* a lot of people will get irate with you, even though that's what the word truly means. *Astronomer* means "star namer." *Astrologist* means "star scientist."

We're all using the words astronomy and astrology incorrectly, and the astronomers want you to use them incorrectly so that no one confuses them with a fringe pseudo-science by the same name. But that's not a technically accurate approach.

Astronomers are notorious for using the wrong words for things. Flat-Earthers take note: Astronomers generally refer to the planets and the Sun as "disks." No kidding. Everyone knows the planets and the Sun are spheres (bumpy, lumpy and slightly oblate spheres, but generally spherical). They are definitely not disks. Why do astronomers do this? Because that's what a planet *looks* like when viewed through a telescope or on a photographic plate. Their primary concern is how to find it in the sky.

Astronomers are (generally) not planetary scientists. They are *orbital* scientists. They track and plot and measure celestial masses and their gravitational fields, and they use the brightness and distance and diameter of the "disk" of light (as seen through a telescope) to identify the object.

But, come on. They know better. We have good, proven definitions for disks and spheres and all the other geometric shapes, even for the ellipses that astronomers use to describe orbits. They know this stuff. And we know that they know it.

If they can casually call a sphere a disk just because it looks like one on a photographic plate, why can't we use the word planet to describe things that look like planets and have all the properties of planets?

For all those kids who have hermit crabs for pets (is that still a thing?), someone should tell them hermit crabs aren't crabs. Neither are coconut crabs. Both are as genetically distant from true crabs as humans are from lemurs. We call them 'crabs' because they *look* like crabs. Biologists are not hindered nor bothered by that.

Also, what you think of as penguins are not true penguins. There once was a bird known as the Great Auk. This bird lived in the far northern arctic and managed to survive for millions of years among the polar bears and arctic foxes. Sadly, the Great Auk went extinct in the 19th century due to over-hunting by humans for their insulating coats of feathers. Its scientific name was *Pinguinus Impennis*. They were also commonly called Penguins.

Years after the last of these unique animals was killed, the sailors who discovered Antarctica saw a similar-looking bird on that remote continent and called them penguins too. But they are not Pinguinus Impennis. They are a distinct and separate species from the true, but extinct, species of penguins.

How the biologists of the world can stand this glaring misclassification is hard to comprehend. They must have stronger stomachs than the astronomical community. To carry on as if the family of *Spheniscidae*, from the order of *Sphenisciformes,* actually deserves the name of Penguin is truly an inspiring example of emotional maturity.

To a biologist, buckwheat is not a type of wheat. It's not part of the wheat family at all. People with wheat allergies or celiac disease can eat all the buckwheat they please without any trouble (assuming it is free of cross-contamination from shared equipment with wheat). Do biologists care that people still eat buckwheat pancakes? Would we all be better off calling it by its scientific moniker, fagopyrum esculentum pancakes?

Bravo, Biologists, for your willingness to accept us as we are, including our casual and inaccurate naming of things. We love you for that.

How about an example from the mineral world?

Rare-earth elements are essential to modern technologies. Your cell phone wouldn't work without them. Fortunately, they're not rare and they're not unique to Earth either.

You're probably familiar with Neodymium magnets. That's one. And Cerium is the 25th most abundant element on Earth; it's more abundant than copper. We call them rare because they don't form in clumps or veins like gold and copper do. They are dispersed and must be filtered and refined out of ores, but they are not rare.

The world is full of words we don't use according to their professional or scientific meanings. But there's no crime in using them the way we do, as long as we all understand the *ideas* being communicated. That's what words are for. They are symbolic representations used to communicate ideas with each other.

What is our common idea for a planet? Is it an object, or an orbit?

Before we get to that, let's look at something more familiar and close to home. Continents and planets are similar in that they are both types of real estate. What do you think the definition of a continent is? Who created that definition? Are we even using the word correctly?

Chapter Four:
A Continental Divide

What is s planet? Maybe we can gain some insight by examining the question of what a continent is. Is Greenland a continent? If not, what about Australia? Why one and not the other? Is Europe a true continent? What about the "sub continent" of India?

If you thought there was a good, technical definition for a continent, I have good news and bad news. The good news is, yes, there is one. Actually, there are many. The bad news is that no one uses them.

Believe it or not, there is still a lot of controversy around these ideas. I found close to a dozen different 'definitions' for a continent from various organizations with official sounding names. One from a school of geology, others by well meaning people with credentials, others not. Most of these definitions referenced a mix of details such as tectonic plates, whether unique plant and animal species are contained within the bounds, the existence of distinct human cultures, a separation by oceans, size, etc...

For example:

Continent:

1. A continent is a very large area of land, such as Africa or Asia, that consists of several countries.

— https://www.collinsdictionary.com/dictionary/english/continent

That's not very specific. Also, Australia contains only one country, so it must not be a continent. Antarctica has no countries, so maybe it's not a continent either. Let's try another one.

A Continent is a very large land mass that is separated from other continents in the world by oceans.

— https://www.tutorialspoint.com/what-is-the-difference-between-an-island-and-the-continent

That seems to rule out Europe as a continent. Other people agree with this. But, that complicates things for Africa (connected to Asia) and also North & South America.

If we take this definition to its logical conclusion, there are five continents — Eurasiafrica, America, Antarctica, Australia and Greenland.

By the way, that's not my idea, and it's not a new idea. According to the World Atlas:

...many people in Russia, the rest of Eastern Europe, and Japan consider Europe and Asia to be one continent, known as Eurasia. In some countries, North and South America are considered one continent, while Europe and Asia are divided. There are even some who believe that Europe, Asia, and Africa should be considered one united continent because they are all joined together by land.

There is also a widely accepted view of what a continent is. This view defines a continent as a large, continuous, distinct landmass, preferably separated by a vast expanse of water. This definition is problematic because many of today's continents are not separated by vast expanses of water. In fact, all the continents are connected by land to at least one other continent, with one exception: Australia.

— https://www.worldatlas.com/articles/is-australia-a-country-or-a-continent.html

Um, Antarctica is connected by land to what, exactly? Is this more evidence that Antarctica doesn't qualify as a continent? Let's check another source.

... there are several largely accepted factors that classify different regions of the world as continents. These factors include tectonic independence from other continents, unique flora and fauna, cultural uniqueness, and local belief in continental status.

— http://www.todayifoundout.com/index.php/2013/04/why-greenland-is-an-island-and-australia-is-a-continent/

That sounds like Antarctica definitely *is* a continent. But the "local belief" part can be problematic. If enough people in Greenland believe it's a continent, is it? If so, what about Hawaii? It fits all the other requirements in that list—tectonic independence, unique flora and fauna, cultural uniqueness... Is Hawaii a continent then?

The "local belief" clause was most likely included for the sake of Europe. Can we find other support to get Europe back on the list?

Europe is generally accorded the status of a full continent because of its great physical size and the weight of its history and traditions.

— https://en.Wikipedia.org/wiki/Europe (taken May 1, 2021)

Apparently the history and traditions of a word should carry some weight behind how it is used. (There certainly is a lot of history and tradition behind using the word *planet* to describe Pluto.)

Let's see what an actual geologist has to say about continents. Doctor of Geology and Youngstown State University Professor Craig Campbell reports:

If you define 'continent' as a large land mass then Europe is not one. It does not have a clear boundary (or break) to the east, politically or culturally. Saying that the Urals are the break is a cop out - there is a lot of territory west of those mountains that is not really like Europe. Neither does any river supply a clear boundary.

I avoid the use of the word 'continent' in my geography classes - it's packed with too much cultural and historical complexity.

— Dr. Craig Campbell, Youngstown State University

https://www.quora.com/Why-is-Europe-considered-a-continent-when-it-is-a-peninsula-of-Asia

Mr. Campbell is telling us that the science of geology doesn't even need the word continent. Later we'll look into whether astronomers can get by without the word planet. There may be better alternatives.

India sits on its own tectonic plate, so, like Australia, it can be thought of as its own continent by some definitions. But Europe is *not* on its own tectonic plate, so it's not a continent. Greenland is large, has distinct species and cultures, and is surrounded by ocean, so by some definitions it should be a continent—except, then Madagascar might also qualify. And as long as we're in that region, Africa is currently in the process of ripping apart at the seams (along the Great Rift Valley) to create a new tectonic plate. Shouldn't it be considered two continents, like Eurasia is, even though Eurasia is only a single tectonic plate?

One of the reasons given for Greenland not being a continent is that it's too sparsely populated. By that logic, Antarctica doesn't qualify either. This is a good example of "cherry-picking" specific details solely for the sake of exclusion, like having an orbit that's slightly unconventional.

It was recently discovered that New Zealand sits on its own private slice of continental crust. How ironic is it that tiny New Zealand can be technically a continent, but since Greenland is part of the North American plate, it's technically only an island.

All this confusion and ambiguity, and yet, the geologists of the world still manage to do their research just fine. No one's pummeling anyone for calling Australia a continent, or Antarctica, or Europe, or South America, even though by one or another definition they can all be excluded from that category. Does anyone really get upset by all this? I'm sure some people do, but what purpose does it serve? It changes absolutely nothing about how actual geologists and geographers do their actual work.

Are the geologists and geographers of the English speaking world going to get all uppity and redefine what a continent is, or isn't, then demote Europe to the status of quasi-continent and expect everyone to fall in line? It seems unlikely. It also seems completely pointless. The list of continents is more about tradition and *general* agreement than any kind of rigorous definition.

I'm sure more than a few people will argue 'but that's not scientific.' Meanwhile, geology is still a hard science that is unhindered by the common use of the word 'continent'. Whether something carries the label of continent or not simply does not matter.

The important classifications are more specific, more granular, and far more useful: arid, frozen, wetland, altitude, latitude, etc… these are the qualities that truly matter. Whether we call it a continent, or a planet, or not, adds exactly *zero value* to the science.

So, if there is no good working definition for a continent, then how does the dictionary handle it? Let's check.

Oh—which dictionary? They're not all the same.

Merriam-Webster defines a continent as: "One of the six or seven great divisions of land on the globe."

Dictionary.com says: "One of the main landmasses of the globe, usually reckoned as seven in number."

My trusty American Heritage dictionary defines it as: "One of the principal land masses of Earth, usually regarded as including Africa, Antarctica, Asia, Australia, Europe, North America and South America."

Notice the weasel words— *usually, six or seven.* There is a definite lack of precision here.

There is often a second definition for continent that's curious. The second definition is usually some variation of: "Europe, as distinguished from the British Isles." This may be the real reason Europe is a separate continent from Asia—it has its own individual, specific definition as such. No other continent gets this treatment.

If we were to treat Pluto this way, we could simply insert a new, second definition for planet: "Pluto, as distinct from the other Kuiper belt objects in its vicinity."

Another fascinating entry for a continent is the second definition at Dictionary.com: "A comparable land mass on another planet." Ah yes, those fabulous continents of Jupiter spring to mind, or Mars for that matter.

Just in case I've confused anyone—this is pure sarcasm. There are no continents on those planets. Come to think of it, the only other planetary body we know of that *might* have oceans, and therefore continents, is Saturn's moon Titan—the one with the thick atmosphere and seas of liquid methane. Except, that's not a planet. It's a moon orbiting a planet. And so, the "on another planet" part of that definition doesn't work. At least, if the IAU has its way.

This highlights another common complaint with the IAU's definition. It requires an object to be in orbit around the Sun to be a planet. Specifically, around *our* Sun. This makes exoplanets not qualify as planets, and therefore they cannot have continents according to the dictionary.com definition of continents. Their definition is useless if we intend to take it and the IAU's definition both at hard face value. This is just the tip of the iceberg of problems we get from being so pedantic about definitions.

Could one of them be (*gasp*) wrong? Maybe both? And why do the various dictionaries differ? Shouldn't the definitions of words be exact and universal? Maybe the history of the word will be helpful.

The word Continent used to mean "held together." Thus, incontinent means "not held together." The word 'incontinent' is still used this way for people who have trouble holding their bowels together. Continent and contained have the same root meaning. Continent is just a nounified conjugation of the word *contain.*

So what is contained within the "continents" of Earth?

It is nothing more than the *idea* of a historically significant division of land. That's it. As was said before, this is what all words are. They represent a common understanding for an idea. The word *common* is critical here. A few hundred people cannot change the common understanding of an idea.

Is sugar classified as a food, or a toxin, or a drug? You can find thousands of people to represent any one of those conclusions, including many scientists. But none of it changes our common idea for how to identify sugar when we see it.

The definition for a word is a description of the most common ideas being represented by it. This is the science of lexicology.

Chapter Five: The Science of Lexicology

Lexicology is just as valid of a science as astronomy or entomology or sociology are, among others. If we're being intellectually honest, we must treat the science of words, how they are used and their definitions, with the same respect and deference as some people insist we do with the IAU's internal pronouncements.

Lexicology is profoundly important to all human endeavors. It is the study of words and how we use them, what they mean, where they came from, how they are related, and also how the words we use shape us as a society of individuals. Lexicographers are the people who use all that information to compile dictionaries.

> **LEXICOGRAPHER, n. A pestilent fellow who, under the pretense of recording some particular stage in the development of a language, does what he can to arrest its growth, stiffen its flexibility and mechanize its methods.**
>
> — Ambrose Bierce, The Devil's Dictionary, 1911.

> **I am not yet so lost in lexicography, as to forget that words are the daughters of earth and that things are the sons of heaven.**
>
> — Samuel Johnson (1709-1784)

What these venerable lexicographers are telling us is that words are transmutable and evolving, and that the things they describe are not changed by that. Shakespeare said it better and more succinctly. "A rose, by any other name, would smell as sweet." But then, he was the slam-dunk grand-master of metaphor.

Why don't the makers of dictionaries just use the definitions that professional organizations tell them to? Because, as in the case with the definition for a continent, how do we decide which group of professionals to grant that privilege to? What's the criteria for that decision and who determines this criteria? How does that person gain such power, and is there an agenda behind their definitions? What should we do if their logic is profoundly flawed? One solution is to allow for multiple definitions.

If you open a dictionary and pick a word at random, most of them will have multiple definitions. Let's use the aforementioned word *argument* as an example. The organization of literary writers uses the word differently than the various organizations of philosophers. And they both use the word differently than computer programmers or astronomers do. In this case, the word 'argument' needs at least four definition for multiple use cases.

> Literary writers: **"Lisa and her father's *argument* over her fashion choices could be heard by the neighbors halfway down the block."**
>
> Philosophers: **"Descartes's *argument* for the existence of free will is fundamentally inert."**
>
> Programmers: **"Supply the function with three *arguments* for frictional coefficient, surface area and pressure to calculate the force required."**
>
> Astronomers: **"The *argument* of periapsis is strictly undefined for objects in an equatorial orbit."**

But which of these groups is more authoritative? Which is more popular? Which is more logical? How do we resolve the problem if one group refuses to bend to another one's will? If we decide to use all four definitions, which one should be first? How do we know there's not a fifth one, or more? Lexicographers determine all this through an exhaustive study of how we use the words in our actual communications.

Is there a dictionary that describes Pluto as a planet? In fact, most of them do. In California and Arizona, Pluto is legally defined to be a planet. There are far, *far* more residents of California, or of Arizona, than there are members of the IAU. That should count for something. Many other countries of the world still refer to Pluto as a planet too, adding to the weight of public opinion. And they have every right to do so. No single organization owns the word.

More importantly, the majority of planetary scientists put Pluto, and Eris and Ceres solidly within the category of planets. And, as the pinnacle experts in the field, they have every right to do so. Their opinions deserves just as much respect and deference as the astronomers' opinions.

Does Pluto, the object, belong to the same category of things as Jupiter and Venus? Do Saturn and Mercury even belong in the same category of things with each other? It depends entirely on how we draw the boundaries for the categories, and whether or not a meaningful category even exists. We'll get to all that soon enough. But first, it is helpful to look at where we came from to get a better understanding of the path ahead.

Chapter Six:
A Nice Little Planet

Before we get to the word 'planet', let's take a quick peek at one of the most popular examples of a word who's definition was radically changed. More specifically, *how and why* it changed.

Nice: a case study

Remember those mealy times back in the 15th century? Life was practical. People worked hard, and 'nice' was a word that all educated people knew came from the latin *ne-scius*, meaning 'not knowing', or to be blunt, *stupid.* But all that was about to change.

The shift in meaning happened when a certain underprivileged class of people developed a habit of overdressing for various situations. Why they were overdressing, we can only guess. Maybe it was an attempt to evade doing hard work by appearing to be in charge, or to nefariously slip into a better social circles than they might normally belong to.

Whatever the reason, it must have been a transparently superficial attempt. These opulent dressers were referred to as 'nice', meaning, of course, stupid.

But lets not forget who the true creators of linguistic change are. Adults merely use the words they already know. Children learn the meanings of words through observation. They adapt the meanings according to how they see a word being used. They also tend to be utterly ignorant of the history and culture behind why it's used that way. To them, 'nice' became a word associated with being dapper and exceedingly polite.

And thus, the meaning changed.

But what in blazes does any of this have to do with Pluto the Planet?

By declaring that "a dwarf planet is not a planet", the IAU has already lost the battle. Kids will see and hear the word *planet*, will hear the teacher say "not a planet", and they will draw their own conclusions about planets and the intellectual integrity of their teachers. And that will be that. Words are defined by how people use them. It's that simple.

Now, hold on a minute. Before the Pluto-is-a-planet party starts the festivities and the Pluto-is-not-a-planet camp gets too excited, let's see if we can offer some deeper insights about this singularly troubled word and help the IAU with its formidable task of classifying objects in the Solar System.

Let's not forget the goal—we don't want to thwart anyone or retard their efforts in the pursuit of truth. The aim of science is to increase the sum total of human knowledge and make that knowledge more useful. That's really what this book is about. We most definitely want science to succeed! After all, science is progress. Let's see if we can be helpful. To start, let's review a little history around the word 'planet'.

Originally there was only one known planet. It wasn't Earth, it was Venus. We know this because of the Babylonians. They were the first to leave any written records for anything that can be called astronomy.

It is widely assumed that knowledge of the planets is prehistoric. But from all the evidence we can find, knowledge of the planets came *after* the advent of writing, accounting, and codified legal systems. You could argue that Stonehenge, or the Pyramids of Egypt, or the Chaco Canyon structures are a kind of record for astronomical observations, but these sites only track the backdrop of stars to align with the spin of the Earth. In some cases they track the Sun and Moon too. That's not astronomy.

The discovery that planets move relative to the stars may have been known in pre-history, but it couldn't have been very important to them. None of the planets appear in any religion's creation stories. The first written reference to a planet comes from the Babylonians, around 1600 BCE, give or take a few hundred years.

The cuneiform Venus Tablet is the first written record of any planetary observations. It only contains references to Venus. No other planets are included. It was written during the reign of King Ammisaduqa, about a hundred years after Hammurabi was king of Babylon. (The Code of Hammurabi is considered one of the earliest written legal systems.)

About a thousand years after the Venus Tablet was written, around 600 BCE, Babylonian culture produced the first known 'astronomical diaries.' This is the oldest known reference to the planets Mercury, Mars, Jupiter, and Saturn. Now there were exactly five known planets in the heavens.

The ancient Greeks learned everything they knew about the planets from the Babylonians, including the word itself. The Greek word, *planetos* meant wandering star, which is exactly what the planets were to them. To the mind of an ancient observer, the dots that moved across the nighttime sky were no different from the dots that didn't move, other than the fact that they moved. Aside from brightness and position, one dot of light looked pretty much like any other dot of light. It was movement, and movement alone, that defined their uniqueness.

The Ancient Greeks divided the sky into three classes: *Asteres Aplaneis*, the "plain stars" that were motionless relative to each other, and *Asteres Planetai*, the "moving stars", plus the Sun and Moon as separate class of objects.

The Romans added the Sun and the Moon to the list of planets in spite of their obvious and extreme differences from the others. This demonstrates that there were problems with the word planet right from the beginning. If it was part of the eternal sky and it moved, to the Romans, it was a planet.

Today, the IAU's definition continues this movement-oriented tradition. Two of the three requirements for planethood are based on its path rather than what we know about the object itself.

In light of modern knowledge, this is a flawed tactic. The idea of classifying items based on the journey they take and the quality of those neighborhoods is similar to a taxonomy of animals based on their migratory paths and the things along that path—like calling a dolphin a fish simply because it swims in the ocean, which the ancients also did.

Of course, the ancients knew absolutely nothing about the quality of the objects in the sky. Defining things by the path they took—the only thing they could know about it—was perfectly logical for them. ***We don't have that excuse.*** We know what the planets look like, up-close and personal. We know their surface topography and the composition of their atmospheres. We have named their mountains, valleys, plains and craters. We have measured their magnetic fields. We know how much they weigh.

Imagine wildebeests and monarch butterflies in the same category simply because they migrate across a continent. Imagine butterflies and beetles not. Humanity eventually learned that dolphins are not fish, and also that the Sun and Moon are not like the other planets, but the Earth *is*. This required changing our fundamental understanding of what those things are.

To reclassify the Sun and Moon and remove them from the list of planets was a big jump. It required a terrible sacrifice. A beautiful and elegant thing had to die.

The ancients believed that all of the heavens revolved around the Earth, but the planets were somehow detached from the heavenly sphere of the stars. It was assumed they each had their own spheres, but the motion of these spheres was weird. They sped up. They slowed down. They passed each other. No one could figure it out.

In a time before the internet, before television and radio, before books and writing were commonplace, this celestial show was an immense source of intrigue and wonder. It occupied people's thoughts, and gave rise to some of the brightest minds in history. This world-view of an Earth-centered system peaked with a Greek-educated Roman citizen named Claudius Ptolemy.

See, there was a big, big problem with the Earth-centered model of the solar system. The motions of the planets were too complex. Unlike the Sun and Moon, some of the lesser planets could stop and reverse their direction, a phenomenon known as retrograde. But the spheres of the Sun and Moon do not. Why?

Ptolemy usually gets the credit for solving this, but there is evidence that others solved it long before him. The Antikythera mechanism does it, and it predates Ptolemy by a couple of centuries. Both the Antikythera mechanism and Ptolemy use a series of circles, epicycles, 'eccentric deferents' and an imaginary point called an equant to create the most accurate geometrical model of the solar system yet devised. The Ptolemaic system was beautiful and elegant, and it would prove to be the most useful model of the Solar System—-until a man named Albert Einstein changed everything.

You were probably expecting that name to be Copernicus. That's what is usually taught in school. But it's not exactly how things happened. The Ptolemaic system for plotting the course of the planets was so accurate, it was used by planetariums, astrologers and some astronomers to calculate the positions of the planets long after the Copernican system was confirmed and widely accepted as true.

That's because Copernicus's model also had its own problem; it wasn't as accurate as the Ptolemaic system. Ptolemy's system was a complicated kludge. But it was an *accurate* kludge. It was the most accurate model of the solar system for well over a thousand years.

This didn't make the underlying assumptions correct. It only made the wrong assumptions useful.

Copernicus generally gets the credit for proving the Ptolemaic system was wrong, and that's close enough to the truth. There were many others before him who 'just knew' that the Sun was at the center. One of these men was Pythagorus (who apparently did not create the theorem named after him, and who did believe that flatulence from eating beans would lead the bean eater to an early death. Sometimes 'just knowing' something isn't good enough.)

Copernicus gets the credit for our heliocentric system because he was the first one to show his math. Even though the Copernican system was more true to nature, it was still less accurate than the Ptolemaic system. People accepted that the Ptolemaic system was wrong, but they continued to use it well into the 20th century. (More about this in the next chapter.)

The Copernican system still didn't tell us anything about the quality of the planets, only their orbits, so the practice of defining planets by their paths continued. The only change was that now a planet was a thing that orbited the Sun.

This solved the problem of why the planets' movements were so different from the Sun and the Moon. It also removed the Sun and the Moon from the list of planets, and more importantly, added the Earth.

The number of planets went from five under the Babylonians, to seven under the Romans, to six (Uranus and Neptune had not been discovered yet.)

Today we are trying to improve our astronomical taxonomy by adding or adjusting minor details—kludges—specifically to include or exclude things that some people 'just know' shouldn't belong. But, like the Ptolemaic model of the Solar System, sometimes it's necessary to rethink our original assumptions and cast them out entirely. Sometimes a system must be replaced by a completely new, fundamentally different system.

In modern times we have seen the planets up close. We have photographed their backsides—a view that cannot be seen from Earth. We have measured them, weighed them, surveyed them, touched them, sampled them and more. Their motion alone is no longer sufficient to describe them.

Telescopes validated Copernicus and put any lingering faith in Ptolemy to rest. Telescopes are still vitally important to the science of astronomy, but we have a new tool now. Modern spacefaring probes allow us to know more details about the quality of the objects in the sky than could be imagined a hundred years ago. And like telescopes, they will usher in a new era of understanding about the order and organization of the Solar System.

Copernicus's model of the Solar System replaced Ptolemy's, but it was never as accurate as the old model until Albert Einstein did a little thing. As promised, here's that story.

Chapter Seven: Albert Einstein, Planet Killer

I promise this is about Einstein, and Pluto. But first, we need to talk about a pointy-eared alien. Say the word 'Vulcan' and most people think of a certain logic-loving science officer from the Starship Enterprise. But Gene Roddenberry didn't invent Spock's world out of whole cloth.

About a hundred and twenty years before the TV show Star Trek first aired, Vulcan was known to be a planet within our own solar system. If you're wondering where it was and what happened to it, well, Albert Einstein killed it.

See, that's how big of a deal relativity is—it's basically the only thing he's known for. Few people remember that Einstein killed a planet, or invented a refrigerator with no moving parts, or helped create a new kind of matter… Yeah, he did a lot more than just revolutionize all of modern physics.

But this story starts with a different hero of science and how he "discovered" Vulcan in the first place. In 1846, astronomer and mathematician Urbain Jean Joseph LeVerrier accomplished a truly amazing feat of scientific wizardry. By studying a strange wobble in the orbit of Uranus and applying some incredibly tedious calculations, he successfully predicted the location of a previously unknown planet, the planet we know as Neptune. He was the first person in the history of the world to discover a planet using pen and paper.

LeVerrier earned an enormous reputation from this success, and the pressure to repeat the trick was too much to ignore. Astronomers knew about an odd wobble in Mercury's orbit too, but no one could explain it. LeVerrier took-on that challenge too. Although his calculations didn't quite work out the same as they did for Neptune, he boldly predicted the location for another new planet in between the orbit of Mercury and the Sun. He named this planet Vulcan, the god of fire and the namesake for volcanoes.

The scientific community's confidence in LeVerrier was so great that Vulcan was added to the official list of planets. Astronomers trained their telescopes as close to the Sun as they dared, searching for the new planet. Except, no one could find it.

To be fair, it is very difficult to see things close to the Sun. Point your telescope a fraction of a degree in the wrong direction and you either miss your target or damage your equipment (and eyeballs) from the intense sunlight. Some astronomers thought they had seen the new planet during the eclipse of 1878, but no one could confirm it. They kept looking.

Then, in 1915, Albert Einstein published his theory of relativity. This replaced the old Newtonian equations for gravity. With his more accurate equations, Einstein was able to prove Mercury was behaving exactly as it should. LeVerrier's math was wrong. There was no other planet disturbing Mercury's orbit.

This is why the Copernican heliocentric system was less accurate than Ptolemy's geocentric system. Newton's equations for gravity aren't quite good enough for celestial mechanics.

In time, as confidence in Einstein's work increased, Vulcan was removed from the list of planets, and the Copernican heliocentric model of the solar system was finally more accurate than the old Ptolemaic system.

But since Einstein's math was so complicated, the Ptolemaic system was still used in planetariums until the 1980s, when new computerized equipment could finally be used to display the relativistic motion of planets. The Ptolemaic system was finally put to rest for good.

But that's not quite the end of the story—

LeVerrier had started a new trend among astronomers. To augment their telescopes and winnow-down the vast quantity of sky to search, more and more astronomers started using math to predict the locations of unknown planets.

There were some irregularities in Neptune's orbit too. A wealthy astronomer named Percival Lowell concluded there must be a "Planet X" out there perturbing Neptune. He built a whole observatory just to find it.

Alas, he never found it. He died in 1916. But the observatory he built in the high mountains of Arizona lived on. His project lived on too. Einstein's math didn't solve the riddle of the wobble in Neptune's orbit. Something was definitely out there.

The observatory hired a young and ambitious astronomer named Clyde Tombaugh to continue the search that Lowell had begun. After many years of effort he ultimately succeeded. And thus, Pluto became the official ninth planet in orbit around the Sun.

Except, it wasn't in the right place to explain the wobble in Neptune's orbit either. New theories about a tenth planet started to spread. People searched, but nothing was found.

The mystery was finally solved in 1989. A little robotic explorer named Voyager-2 was sent out to take a closer look. It discovered that we had Neptune's mass wrong. Armed with the correct value for its mass, new calculations showed that Neptune was doing just as it should. No perturbing body was necessary to explain its behavior.

Sometimes all you need is a closer look.

Chapter Eight:
The Starship Transcendent

"Captain's log, galactic standard date 2727.359. The Starship Transcendent is responding to a distress call from the seventh planet in the Traaub-Farka system. The nature of the distress is unclear. We will enter orbit around the planet in a few moments, where we will render whatever aid and assistance we can."

"Helmsman," the captain said, "Put us in a low orbit around the seventh planet."

"I'm sorry sir, I can't do that. The seventh planet is a highly radioactive gas giant. We can't get too close."

The captain's knitted his brow at the misunderstanding. "No, not the gas giant, that yellow one." He pointed at view screen at the front of the bridge, indicating the planet just inside the gas giant's orbit. "Isn't that the seventh planet?"

Before the helmsman could respond, the science officer answered, "Captain, I believe that is a dwarf planet."

"A dwarf planet? Are you sure? It's got to be twice the size of Earth."

"There is too much debris in its orbit. It fits the definition."

The captain studied the view screen again. "Computer," he said, "Show a wide view of this system. Display orbital clutter."

The screen zoomed out and filled with wispy clouds of orange and yellow. The colors were brighter where the concentrations of debris was higher. There were two large clumps on either side of the gas giant, and several other bands of color in other parts of the system.

"What about all the debris in the gas giant's orbit? There seems to be a lot more debris there," the captain said.

"Gas giants are exempt from the debris rule due to the *Margot Mass Ratio* equation."

"Well, whatever it is, put us in a low orbit around that yellow planet."

The science officer turned his attention to the helmsman and made eye contact. The helmsman recognized the silent message. He reported to the captain, "No can do sir. Again, it's a dwarf planet. Low orbits are against regulations for dwarf planets."

"Why is that?" the captain demanded, testing his subordinate.

"Low planetary mass rule. When we leave orbit there's a risk the ship's warp field could nudge a dwarf planet to a different orbit."

Agitated, the captain thrust his hands toward the screen. "Look at it helmsman. That's obviously not a low-mass planet."

The science officer interrupted. "Mass is not part of the definition for a Dwarf Planet. It's only given as the reason we can't make a low orbit around one. The regulation is separate from the definition."

"Well, ignore the regulations and put us into orbit."

The helmsman hesitated. "According to regulations, I'll need confirmation from a ranking officer that the captain is sane and healthy, and his orders are not putting the ship or any local populations in danger."

——— *Several tedious minutes later* ———

"Captain," the helmsman said.

"Yes helmsman."

"We are entering orbit around the dwarf planet now." He checked his console, then added, "but it seems to be uninhabited and there's no sign of a distress call."

"Explain, ensign. This is the seventh planet, isn't it?"

"Technically, no. But if we include all the larger planet-like objects in the count, then this is the seventh planet-like object from their sun."

The captain turned to the communications officer. "Lieutenant, do you have a bearing on the distress signal?"

She double-checked the data on her screen. "Yes captain, but it's coming from the outer regions of the solar system."

"Can you pinpoint the location?"

"No sir. They seem to be using a system of relays. The signal is being repeated from hundreds of locations distributed among the scattered disk. It's the best way to distribute a signal, but impossible to tell where the signal originated without knowing their transponder codes."

"Locations? Do you mean asteroids? Are any of those relays on something we could call a planet?"

Before the communication officer could answer, the science officer interrupted again, "They could be asteroids, or comets, centaurs, trojans, plutoids, sednoids, cubewanos, etcetera… we don't know enough about this system to make that determination."

Frustrated, the captain raised his voice and spoke to the room, "Analysis? Does anyone have any useful information?"

The science officer responded, "This system is less than a billion years old. There hasn't been enough time for planets to sweep-up all the debris of formation. Many of the objects here may look like planets, but they could easily be dwarf planets."

"But this 'dwarf planet' is bigger than Earth!" the captain exclaimed.

"Yes, that's true. But until we do a survey of all the mass in the system and determine what's resonating with what, we don't know which of the objects here are a true planets."

"Resonating?" The captain's voice resonated with growing impatience.

"Yes sir," the science officer responded, "It's not enough to know how many objects there are or what their mass is. We also need to know the shape and size of their orbits, and if they're in a co-orbital resonance with another object, or follow or lead an object in its Langrangian points like the trojans in Jupiter's orbit, or the plutinos in harmony with Neptune. Those objects don't count, regardless of their size."

The captain fumed. "Get to the point, mister. People are suffering. How long will it take to figure out which is the seventh planet?"

"I'll have to survey the solar system, formulate a set of tensor field equations, integrate the results and… I can have an answer within the next two or three days."

"Two days!"

"Yes sir. It takes time to track the path of an orbit. There are a lot of orbits in this system."

"Captain," the helmsman interrupted. "If we ignore regulations on intra-system warp-speed limits, we can visit every planet…" He stopped to glance at the science officer. "Excuse me sir, we can visit every object of significant size in the system within a few hours."

The captain turned to the communications officer. “How fast can you detect a local signal once we get to a planet,” he hesitated, gave the science officer a sideways glance, then continued, “or a significantly sized object?”

“I only need one pass around the object,” she said.

“Very well. Turn on the sirens helmsman. Let’s go help those people. They’re obviously astronomers, so they can’t have any survival skills to speak of.”

The science officer spoke, “I’m curious captain, how did you come to that conclusion?”

“Any practical person asking for help would have used a more obvious determinant for what a planet is. Only an astronomer would care more about the orbit of a thing than the thing itself.”

The science officer cocked his head. “But, that’s not scientific.”

“It’s language, lieutenant. Communication shouldn’t require measurements and analysis just to get an idea across. Science is important, but orbital dynamics are not relevant to the objective of this mission. And it ignores the most important part of what a planet is—whether or not people want to explore it. We may use the same words, but context demands variations on the meaning. Now let’s go save those people.”

Section Two

Science and Data

Chapter Nine: Scientific Definitions

You can't do science without data. Let's look at some other scientific definitions to get an idea for how these things work. Scientific constants like kilograms and meters have arbitrary definitions, but scientific constants like the speed of light and pi are first discovered, then defined based on measurements. We've talked about continents, but what's the scientific definition for an ocean, or a sea? Where does the Atlantic Ocean stop and the Indian Ocean start?

...the Earth's major landmasses are surrounded by one, continuous World ocean that has been divided into a number of principal 'oceans' by the land masses themselves and various other geographic criteria.
-- https://www.universetoday.com/72611/what-is-a-continent/

It turns out that there is no scientific definition for what is an ocean, other than a vaguely large body of water. This is also true for a sea. The oceanographers and marine biologists of the world don't seem to be bothered nor hindered by this. Meaningful oceanographic science continues to be done.

What about a mountain? What's the difference between a mountain and a dwarf mountain (otherwise known as a "hill")? Or a lake and a dwarf lake (also known as a "pond")? As it turns out, there are several definitions to choose from. And once again, they are of little to no use to anyone in the scientific community.

Between the 1920s and the 1970s, government agencies in both the US and the UK considered anything higher than 1000 feet to be a mountain. But the United States dropped the 1000 foot requirement in 1970 and now has the official position:

There is no technical definition for a mountain.

— USGS website

In the UK, they changed the definition. They increased the measure of a mountain from 1000 feet to 2000 feet. That's not a small adjustment. Also, the round numbers expose how arbitrary the concept is.

There's nothing arbitrary about pi or the speed of light.

The UN Environmental Programme made a more nuanced attempt to clear up the issue. They don't define the word mountain specifically, instead they define a 'mountainous environment' as meeting any one of these conditions:

1. **Elevation of at least 8200 feet (250 meters).**

2. **Elevation of at least 4900 feet (1500 meters), and a slope greater than 2 degrees.**

3. **Elevation of at least 3300 feet (1000 meters), and a slope greater than 5 degrees.**

4. **Elevation of at least 980 feet (30 meters), with a similar elevation range within 4.3 miles (7 km).**

That last one is intended to contain mountain ranges. However, using these criteria, 24% of North America is 'mountainous', and fully one-third of Eurasia qualifies.

—Blyth, S.; Groombridge, B.; Lysenko, I.; Miles, L.; Newton, A.;

2002, Mountain Watch, UNEP World Conservation Monitoring Centere, Cambridge, UK.

You might think all these different uses of the word mountain would cause problems, but the geologists and geographers of the world are unhindered by any of it.

We can't get too nit-picky about specific details of mountain heights without opening up a whole other can of worms around how to measure them. Altitude can be measured from the base of the mountain, or from sea level, or as the distance from the center of the Earth.

Complicating this is the fact that sea level changes with the tides, and the Earth is not exactly round. Because the Earth spins, sea level at the equator is farther from the Earth's core than it is at the polar latitudes.

For this reason, there is more than one "highest mountain" in the world. Mt. Everest is the highest above a calculated average for sea level. But if we measured its height from the actual sea level hour-by-hour, its height changes constantly with the tides, as do other mountain heights at other global positions.

More reliable measures can be found with the other "tallest" mountains.

Mauna Kea in Hawaii is the world's tallest mountain as measured from its base (which happens to be below sea-level). The top of Mt. Chimborazo in Ecuador is usually considered the farthest point from the center of the Earth, even though it is more than 6000 feet closer to sea level than Mt. Everest. The two are on different latitudes, and sea level bulges out at the equator due to the spin of the Earth and the tug of the moon.

So why does Everest get the nod for tallest mountain? Because the local conditions (extreme cold, low oxygen, long arduous trek to get there, etc.) make it *seem* like the obvious choice. Earth's atmosphere also bulges at the equator, which is why it's easier to breathe at the top of Mt. Chimborazo than at the peak of Everest.

That's not a very scientific method for determining the highest mountain, but there is no better explanation. The top of Everest is definitely *not* the point farthest from the center of the Earth, and it is *not* the tallest natural formation on Earth either.

But how does the highest authority in the world for the English language define a mountain? The Oxford English Dictionary says a mountain is:

A natural elevation of the earth surface rising more or less abruptly from the surrounding level and attaining an altitude which, relative to the adjacent elevation, is impressive or notable.

Applying similar logic, Ceres has definitely attained a notable and impressive 'altitude' relative to the other objects in the asteroid belt. The same is true for Pluto, Eris and Sedna <u>within their orbital zones</u>. But that's the end of the list. Those are *by far* the largest objects within their categories of objects, including all of the known trans-neptunian regions.

Maybe mountains and continents are a fluke. Let's look at some other examples. For instance, what is a river?

According to the United States Geological Survey (USGS):

Rivers? Streams? Creeks? They are all names for water flowing on the Earth's surface. As far as our Water Science site is concerned, they are pretty much interchangeable.

— https://www.usgs.gov/special-topic/water-science-school/science/rivers-streams-and-creeks?qt-science_center_objects=0#qt-science_center_objects

Unfortunately, the meteorologists of the world might take issue with the "flowing on the Earth's surface" part of that definition. According to the National Oceanic and Atmospheric Administration (NOAA):

Atmospheric rivers are relatively long, narrow regions of the atmosphere — like rivers in the sky — that transport most of the water vapor outside the tropics.

— https://www.noaa.gov/stories/what-are-atmospheric-rivers

The most well-known atmospheric river is called The Pineapple Express. It brings water from the tropics near Hawaii to the United States' west coast. But land and sky aren't the only place you can find a river. The Gulf Stream is also considered a river of water within the Atlantic Ocean.

All this history and confusion has led the USGS to make a single, comprehensive, <u>official</u> statement on the issue of defining natural formations:

What is the difference between "mountain", "hill", and "peak"; "lake" and "pond"; or "river" and "creek""? There are no official definitions for generic terms as applied to geographic features. Any existing definitions derive from the needs and applications of organizations using those geographic features.

These categories generally match dictionary definitions, but not always. The differences are thematic and highly subjective. For example, a lake is classified by the GNIS [Geographic Names Information System] **as a "natural body of inland water", which is a feature description that can also apply to a reservoir, a pond, or a pool. All "linear flowing bodies of water" are classified as streams under the GNIS. At least 121**

other generic terms fit this broad category, including creeks and rivers. Some might contend that a creek must flow into a river, but such hierarchies do not exist in the nation's namescape.

Broad agreement on such questions is essentially impossible, which is why there are no official feature classification standards.

— https://www.usgs.gov/faqs/what-difference-between-mountain-hill-and-peak-lake-and-pond-or-river-and-creek?qt-news_science_products=0#qt-news_science_products

I like the wisdom of that last part. The statement, "Broad agreement on such questions is essentially impossible," also describes the issue of what a planet is.

In Oklahoma, *Mount* Scott is only 823 feet high, but the Black *Hills* of South Dakota are either 7242 feet high or 2922 feet high, depending on whether you measure from sea level or from the local terrain. And that's here, on Earth, in our own backyards. Did we really think Space was going to be different?

A similar situation exists in Biology. What is a species?

The line is not as clear as most people would like to think. There are a number of competing definitions. Most people tend to think of reproduction as defining a species, but where do asexual organisms fit into that? What about lions and tigers (and others) that are two distinct species according to the phylogenic and polyphasic methods of delineation, but are also still able to reproduce?

This question was studied by anthropology researcher Stefan Milo when discussing whether or not modern humans and neanderthals were genuinely different species, or were the same species but with different morphology. In his research, Milo discovered more than 26 different definitions for species.

Dogs are a good example of a single species with vastly different shapes and sizes. The range of dog morphology is far greater than the differences between modern humans and neanderthals, but all dogs are considered the same species.

A different example is giraffes. They all look similar, but depending on which experts you talk to, there are either four distinct species of giraffe, or there is one species of giraffe with nine sub-species. Each group of experts uses a different criteria for its determination. There is no single, grand, "official" definition for what makes a species.

No one definition has satisfied all naturalists; yet every naturalist knows vaguely what he means when he speaks of a species.
— Charles Darwin, On the Origin of Species.

The line between 'alive' and 'not alive' might seem obvious, but there is no universal definition for determining the quality of being alive. For any definition of life, it is possible to find non-living things that fit. Fire is a great example of something that moves, consumes resources and reproduces. And there are things that most people consider "alive" that don't fit any of the most common definitions of life—viruses, for example.

And the science of biology marches on regardless.

In the sciences of sociology, history and anthropology, the words 'civilized' and 'civilization' carry some measure of weight. But in the book, *Civilization*, author Niall Fergusson dedicates an entire chapter to the task of defining these important words. His conclusion: A hard, scientific definition for civilization is essentially impossible.

What about chemistry and physics?

It turns out that Chemistry has a good working definition for what an element is. It's determined by the exact number of protons in the nucleus of the atom. Physics has a hard working definition for the four fundamental forces and all the SI units we use for measurements. But when both disciplines are pressed on the question of 'what is a metal?', the answer is always, 'it depends.'

In chemistry: "The number is inexact as the boundaries between metals, non-metals, and metalloids fluctuate slightly due to a lack of universally accepted definitions of the categories involved."
— Wikipedia article: Metal, Retrieved 08/16/2021

In physics: A metal is any material (including compounds) that can conduct electricity at a temperature of absolute zero.

In astrophysics: A metal is any element heavier than helium.

Somehow, civilization carries-on without collapsing from this lack of specificity. No one is admonishing school kids for thinking that carbon, oxygen and silicon are not metals, even though this violates the sensibilities of the astrophysicists.

Let's think about that for a moment. The same astronomers who want you to think that Pluto is not a planet, also consider oxygen and nitrogen to be metals. Breathe deep our metallic atmosphere, my friend. (How many bar bets can you win by insisting that you can see through metal?!)

But this ignores the context for how they use the word metal. One definition does not suit all situations.

The vast majority of planetary scientists recognize a subtle truth about the IAU's definition of a planet; it's about as useful as the fuzzy green casserole leftovers from last month growing in the back of someone's fridge. It should be thrown out with the garbage. It is specious definition who's sole purpose was to filter out Pluto from the list of planets and nothing else. In terms of scientific utility, it is utterly worthless. No one should be using it.

When your own people openly refute the standards you create, that's a sure sign that the standards were not created for the good of the community. Something else was driving it.

In any organization there will be favoritism and politics. People will have agendas and nefarious intents. This is normal. It's just human nature. It's obvious this is what happened with the definition for a planet. It is a definition that serves no real purpose. It doesn't need to exist.

As demonstrated in all the other sciences, hard scientific definitions for common words and categories are completely unnecessary. They may even be impossible. This applies equally to the word planet.

That's not just my opinion. It is the opinion of a great many experts.

Chapter Ten: Expert Opinions

The word 'planet' has outlived its usefulness, It doesn't celebrate the scientific richness of the solar system.

— Neal deGrasse Tyson, https://www.space.com/9553-pluto-planet-experts-weigh.html

The IAU definition doesn't really help, and really just complicates things. ... 'Planet' isn't a definable term. That's because it's not a rigidly scientific word. *It's a concept.* The whole idea of a 'planet' is necessarily somewhat vague.

— Phil Plait, the Bad Astronomer https/www.syfy.com/syfywire/redefining-planets-answer-search-question

Scientifically, whether Pluto is also a planet is a non-issue. No scientific definition of a planet exists or is needed. ... Scientists are interested in learning about the origin of the solar system, and setting up arbitrary definitions of planet-hood is of no help here.

— David Jewitt, astronomer at the University of Hawaii.

I think it's a terrible definition.

— David Charbonneau, researcher at the Harvard-Smithsonian Center for Astrophysics.

The assumption is there that the IAU has a corner of the market on what a definition is. We in the planetary science field don't need the IAU definition. The definition of words is based partly on how they are used [...] the definition that people use and what teachers teach, it will become the de facto definition, regardless of how the IAU votes in Prague.

— Kirby Runyon, Dept. of Earth and Planetary Sciences, Johns Hopkins University.

They went too far in demoting Pluto, Way beyond what the mainstream astronomers think.

Most planetary scientists call Pluto a planet and don't [care] what the astronomers say. We are the experts in this field, not the astronomers. We wouldn't pretend to classify galaxies to them because we're not expert in it. I tell public audiences, Don't go to a podiatrist for brain surgery, don't go to an astronomer for planetary science.

If you go to a planetary science meeting [...] you'll hear people calling Pluto a planet and Eris a planet and and the other small bodies in the Kuiiper belt planets, and Ceres...

— Alan Stern, Associate Administrator for Science at NASA, Principal Investigator: New Horizons Mission.

The IAU definition would say that the fundamental object of planetary science, the planet, is supposed to be defined on the basis of a concept that nobody uses in their research.

— Philip Metzger, Planetary Scientist, UCF

Metzger and his colleague Kirby Runyon studied two-hundred years' worth of scientific papers on the subject of planets to discover the actual usage pattern for the word, as any lexicographer worthy of the title would do. This kind of research is the true and legitimate way definitions are made. Their conclusion:

We now have a list of well over 100 recent examples of planetary scientists using the word planet in a way that violates the IAU definition.

They found only one case in the literature that supports the IAU definition. It was published in 1802, a time when star charts were still being drawn by hand and nothing was know about the planets other than their position in the sky and their brightness.

Here are some of their recent examples—

When describing Saturn's largest satellite Titan, Scientist C.P McKay and seven other scientists collectively report:

A planet-wide detached haze layer occurs 300-350 km above the surface; the visible limb of the planet, where the vertical haze optical depth is 0.1, is about 220 km above the surface.

— C.P. McKay et al, 2001, Plan. Space sci., 49

The New Horizons team reported in 2015:

Pluto [is] the largest of a new class of small planets formed in the outer solar system during the ancient era or planetary accretion. ~4.5 billion years ago. ... This suggests other small planets of the Kuiper Belt, such as Eris, Makemake and Haumea, could express similarly complex histories and rival those of terrestrial planets. ... Pluto's diverse surface geology and long-term activity raise fundamental questions about how small planets remain active many billions of years after formation.

— Journal of Science, The Pluto system: Initial results from its exploration by New Horizons. 16 Oct., 2015.

The man who the Kuiper belt is named for said,

[Beyond Pluto may be] one or more small planets, like Ceres

—Gerard Kuiper, *1951, Astrophysics: A topical Symposium, McGraw-Hill Book Co., pg 402.*

(Note that he's referring to Ceres as a planet, in 1951. That's 100 years after it was supposedly reclassified as an asteroid.)

Metzger argues that Pluto and Ceres should be classified as planets, as should the large spherical moons and so-called 'dwarf planets', for a solid scientific reason: Their gravity is sufficient to give them the property of being spherical. He says:

That's not just an arbitrary definition, It turns out this is an important milestone in the evolution of a planetary body, because apparently when it happens, it initiates active geology in the body.

Metzger and Runyon are not alone in their reasoning.

The reason we do the classifications is to try to find patterns that will help us to understand how it works or how it came to be, and if you want to understand how the interiors of solid bodies work, then you should probably be thinking of Pluto as a planet.

— Michael A'Hern, comet and asteroid specialist, University of Maryland, President of IAU's Planetary Systems Sciences Division.

The IAU's definition was created by non-experts—astronomers—who study stars, galaxies and black holes. They botched it.

— Dr. Alan Stern, Planetary Scientist, head of NASA's New Horizons Mission.

Stern also said:

The purpose of the IAU creating this definition was to limit the number of planets in our Solar System so that school kids wouldn't have to memorize long lists of planets.

Okay. But what does an actual science educator have to say?

I favor Pluto, Xena (or whatever its ultimate sobriquet), Sedna, and others being called generally 'planets.' Then we'd have teaching opportunities with the adjectives or descriptors.

— Bill Nye, the science guy.

By using the word *sobriquet,* Mr. Nye was signaling that his opinion should carry a little extra weight. He's not just another casual lay-person. He's one of the team. He knows the professional lingo. The word *sobriquet* usually means a nickname or a common designation. It's a word astronomers often use when naming newly found objects.

In 2017 Stern, Runyon and a group of other planetary scientists issued a paper on the subject of the word planet:

In the mind of the public, the word 'planet' carries a significance lacking in other words used to describe planetary bodies. ... Many members of the public, in our experience, assume that alleged 'non-planets' cease to be interesting enough to warrant scientific exploration. ... To wit: A common question we receive is, 'Why did you send New Horizons to Pluto if it's not a planet anymore?'

They proposed a new working definition for planet:

A planet is a sub-stellar mass body that has never undergone nuclear fusion and that has sufficient self-gravitation to assume a

spheroidal shape adequately described by a triaxial ellipsoid regardless of its orbital parameters.
— Lunar and Planetary Science XLVIII (2017), paper 1148.pdf

They go on…

The eight planets recognized by the IAU are often modified by the adjectives 'terrestrial,' 'giant,' and 'ice giant,' yet no one would state that a giant planet is not a planet. Yet, the IAU does not consider dwarf planets to be planets. We eschew this inconsistency. Thus, dwarf planets and moon planets such as Ceres, Pluto, Charon and Earth's moon are 'full-fledged' planets. This seems especially true in light of these planets' complex geology and physics.
— Lunar and Planetary Science XLVIII (2017), paper 1148.pdf

In other words, all spherical objects in space that have never been stars should qualify as planetary bodies. (Neutron stars with twice the mass of our Sun can be less than 1/10th the diameter of Ceres, and they are nearly perfect spheres, so size and shape alone are not sufficient qualifiers.)

This definition is intuitive. It is based on the object's intrinsic properties rather than its location, or its neighbors, or its trajectory through space. It is also consistent with non-scientific use of the word for any celestial object that might one-day be explored or colonized by people.

With such strong reactions floating around, why did the Astronomical Union feel this definition was necessary in the first place?

It's simple. They were following the 200 year old precedent of Ceres's demotion. It is important to point out that Ceres was demoted from planethood in a time when nothing was known about Ceres or how completely different it is from the other objects around it. Remember, it was described as 'star-like.' We now know it is nothing like a star. None of the 'asteroids' are. If they had been able to see Ceres as we have seen it, in all its roundness and with its brine cryovolcanoes and other geologic features, a different precedent would have been established. This is not conjecture. The man who discovered Ceres said so himself:

Neither the appellation of planets nor that of comets, can with any propriety of language be given to these two stars ... They resemble small stars so much as hardly to be distinguished from them ... From this, their asteroidal appearance, if I take my name, and call them Asteroids; reserving for myself, however, the liberty of changing that name, if another, more expressive to their nature, should occur.

— William Herschel, 1802. On the observations of Ceres and Pallas

If the name 'planet' is not suitable for Pluto, then the name 'asteroid' is certainly not suitable for the small, hard, cold chunks of rock and ice in orbit around the Sun. Besides that, following precent without regard for new information is just bad science.

Scientists have little interest in what has gone before. If new evidence renders a theory untenable, then that theory is rejected and a new one imposed, regardless of how long or how deeply the former theory was believed ... Science is the one discipline in which purging the books of past ideas is not only common practice, but correct.

— ParallaxNick, astro-historian.

There you have it. Real scientists correct the errors of their predecessors. This is exactly what a handful of astronomers wanted to do with Pluto. But as we'll see later, they were "fixing" the wrong problem.

Their reasoning was less than scientific. It was more emotional and political. Rather than correct the errors of the predecessors who thought Ceres was just another asteroid, they used that precedent as leverage in a growing schism within the astronomical community—what is more important, the orbit, or the object? Dynamics, or structures?

Chapter Eleven: Pluto Is Different

The primary reason the IAU says Pluto can't be a planet is because there's an inconvenient Kuiper Belt next door. But the reason reason given for a decision is not always the same as the original reason the decision was made.

A few astronomers have cast doubt on Pluto's planethood since long before the Kuiper belt was discovered, based entirely on the shape of its orbit. Their reasoning was not that Pluto's orbit has other things near it—they didn't know about those things yet. It was because Pluto's orbit is too different from the other planets and asteroids. This attitude that Pluto is "different" continued even after we learned more about Pluto's true nature.

Pluto contains about 70 percent rock and 30 percent ice if you measure things by mass. But rock is denser than ice, leaving the rocky parts to occupy only 45 percent of Pluto's volume. If volume is what matters to you, then you can rightly declare that Pluto is mostly ice. This fact sits within a long list of properties that are not shared with any other planet in the solar system.

— Neal deGrasse Tyson

That is the opening line in a chapter three of Mr. Tyson's book, *The Pluto Files.* But there is a flaw in Mr. Tyson's logic. Someone should tell him that Saturn is even less dense than Pluto, so, you know, they kind-of have that low-density thing in common. Venus is the only object in our solar system known to be entirely devoid of ice. Even Mercury has ice at its poles. Pluto is sometimes called an ice dwarf, while Neptune and Uranus are well known as the ice giants. They all seem to have an ice-thing in common. So Pluto is not so different in that regard either.

Mr. Tyson claims to be neutral on the issue of Pluto's planethood, but the use of deceptive logic like this exposes a different truth. If mass is what's important to you, then Pluto is mostly rock, which Saturn and the other gas giants are definitely *not,* but the other rocky planets *are*. Pluto is far more similar to Earth than Earth is to any of the gas giants.

For example—

It's got nitrogen ice caps. It's got seasons. It's got a moon. It's got an atmosphere. It's got a whole suite of properties which distinguishes it from what we know about any other Kuiper belt object.

— Mark Sykes, Planetary Scientist and chair of the Planetary Sciences Division of the American Astronomical Society

Pluto is round. It orbits the Sun. It has moons. It has an escape velocity way too big for it to become a comet no matter how close it might get to the Sun. It has an atmosphere. It has clouds. It has glaciers. It has volcanoes. It has a solid surface with active geology and mountains and plains that can be landed on and explored. It is profoundly more like a planet than any of the gas giants (that have no surface), or Mercury (that has no atmosphere).

Calling Pluto a planet is linguistically valid and accurate. Anyone who says, "that's not scientific," should issue an apology to all the people who practice the actual science of linguistics.

Neil deGrasse Tyson once said that if Neptune was the size of a Chevy Impala, Pluto would be the size of a matchbox car. Thus, it is too small to be a planet. But as Philip Metzger points out, Mr. Tyson is flat-out wrong. A Chevy Impala is actually 64 times the length of a Matchbox car. Neptune's diameter is only 21 times the diameter of Pluto. Mr. Tyson exaggerated his claim by a factor of three.

Philip Metzger also points out that Jupiter is 95 million times larger than the smallest round object (KBO 2010-GB_{174}). That may sound like a lot, but the largest mammal (blue whale) is 101 million times larger than the smallest mammal (etruscan shrew). The largest fish is 4,000 million times bigger than the smallest, and looking out into space, the largest known galaxy (IC 1101) is 15,300 million times the size of the smallest galaxy (Segue 2). Pluto is the 17th largest object out of the millions of things in our Solar System. It's not that different from Earth by relative size comparison.

Pluto's size gives it significant gravity too. A lot of people think it's just an oversized comet. But it's not. The physics don't work out for it to grow a tail the way comets do. If Pluto dove into the inner solar system all the way to the orbit of Mercury, the sublimating gasses would be held by Pluto's gravity to form a thicker atmosphere. No tail, no comet.

The earliest reasons given for why Pluto should not be a planet were based on the eccentricity and inclination of it's orbit.

Let's take a look at those numbers:

Planet	Orbital Eccentricity	Inclination to Ecliptic (Earth's orbit)	Inclination to Sun's Equator	Inclination to Invariable Plane
Mercury	0.206	7.005	3.38	6.34
Venus	0.007	3.395	3.86	2.19
Earth	0.017	0.000	7.155	1.57
Mars	0.093	1.850	5.65	1.67
Ceres	0.076	10.59	—	9.20
Jupiter	0.048	1.30	6.09	0.32
Saturn	0.054	2.49	5.51	0.93
Uranus	0.047	0.77	6.48	1.02
Neptune	0.009	1.77	6.43	0.72
Pluto	0.244	17.16	11.88	15.55
Eris	0.436	44.04	—	—
Sedna	0.85	11.93	—	—

(Data compiled from various Wikipedia entries)

I included Pluto, Eris and Sedna, but not the other dwarf planets because (as we'll see in the chapter on the Kuiper belt) these are the largest members of different overlapping systems within the trans-neptunian region.

Inclination to the ecliptic is only useful for locating things in the sky from the perspective of Earth. It has little bearing on the dynamics of the Solar System.

The inclination to the Sun's equator is a better metric for understanding the dynamics of the solar system. Notice that the planet with the second-greatest inclination to the Sun's equator (after Pluto) is the Earth!

Measuring the inclination to Jupiter's orbital plane might be more meaningful due to its extreme gravity. Excluding the Sun, Jupiter contains more than half the mass or the entire solar system. A close variation of this is called the invariant plane. It's an average of the angular momentum of all the mass in the Solar System.

It's obvious from this chart that the farther from the Sun we go, the looser the orbits tend to become (more eccentric). This is not surprising since the tug of the Sun's gravity diminishes with distance. But even the closest planet to the Sun can be problematic.

Mercury is the greatest exception to the rule by being the closest planet to the Sun (where orbits tend to be neater and tighter) but also having a wildly eccentric orbit. Does this make Mercury a non-planet? Especially since Mercury is also the only planet without an atmosphere? It seems that Mercury is the true oddball in the Solar System.

There's actually a lot more to the quality of an orbit than just eccentricity and inclination. Wikipedia lists at least a dozen characteristics for an orbit, including Aphelion, Perihelion, Semi-Major Axis, Eccentricity, Orbital Period, Synodic Period, Average Orbital Speed, Mean Anomaly, Inclination, Longitude of Ascending Node, Time of Perihelion, and the Argument of Perihelion. In addition to these, there are orbital resonances, tidal locking and three-body problems when two objects in unrelated orbits get too close to each other.

Can one word really encapsulate all that into a discrete category? That word, *planet*, also needs to encapsulate the physical state of the object. For that, Wikipedia lists no less than twenty characteristics, including: Mean Radius, Equatorial Radius, Polar Radius, Flattening, Surface Area, Volume,

Mass, Mean Density, Surface Gravity, Moment of Inertia Factor, Escape Velocity, Rotation Period, Sidereal Rotation Period, Equatorial Rotation Velocity, Axial Tilt, North Pole Right Ascension, North Pole Declination, Albedo, Surface Temperature, Atmospheric Pressure, Atmospheric composition…

If we regard the planet's axial tilt as part of its orbit, then Uranus is by far the biggest outlier. Its tilt is more than 90 degree from its orbital plane. This profoundly affects its weather patterns and the orbits of its moons.

Perusing these values, one thing becomes abundantly clear. Every planet is vastly different from every other planet in one way or another. Pluto is no exception. Cherry-picking one or two or five specific qualities as a reason to exclude it is disingenuous and deceptive.

As Margaret Mead said,

> **Always remember that you are absolutely unique, just like everyone else.**

This goes for planets too. Considering all this, does the number and quality of its neighbors seem like the best metric for whether something can be called a planet? It wasn't the right thing to do in the case of Ceres, and it's not the right thing to do with Pluto.

The discovery of Eris added more fuel to the fire of whether Pluto should be considered a planet or not. Eris is more massive than Pluto, but also more dense, making it slightly smaller. Instead of declaring Eris as the tenth planet in a new era of discovery (which NASA actually did), a handful of astronomers made the decision that ten planets was just too many planets.

That decision was based on the idea that more Pluto-sized objects *might* eventually be found, maybe even larger than Pluto. It was all very distressing. Who's going to name all these things? What's the criteria for which ones are planets and which ones are not? Who gets to decide?

Chapter Twelve:
The Politics Of The IAU Definition

In a nutshell, the vote they took wasn't logical, it was emotional.

It would be disastrous for astronomy if we come away from the general assembly with nothing. We will be regarded as complete idiots.
— Michael Rowan-Robinson, from the book, *The Case For Pluto*, by Alan Boyle, p 133.

Sorry Mr. Rowan-Robinson, you were wrong. The definition that was fought over and ultimately installed has made the astronomical community look more like idiots than the lack of a definition ever could have.

The obvious logical and honest choice would be to admit that there is no consensus, the proposed definition is problematic, and therefore no definition should be issued at that time. This would have injured or hindered exactly no one in their scientific careers, as proven by literally every other branch of science that marches on without any hard, specific definitions for their most well-known categories.

Declaring that Pluto is *not* a planet actually did hinder and injure quite a lot of people. Specifically, the entire New Horizons team. And it was a hard slap in the face of Clyde Tombaugh's legacy, the dedicated astronomer who spent years of hard and serious effort sifting through a mountain of images until he found that elusive needle in the vast haystack of the cosmos and discovered Pluto. The man deserves much more respect than he gets.

After Pluto, it took sixty two years and several advances in technology to find the next big object in our Solar System. That's only four years less than the time it took between the Wright Brothers' first powered airplane flight and landing a man on the Moon.

The IAU's bastardization of a definition gave all of astronomy a big black eye. It disappointed and alienated school kids and their teachers the world over. It was a very bad move.

Who is the International Astronomical Union (IAU), and why do they exist?

The IAU was founded on July 28, 1919. Improvements in the math and in the tools used to survey the skies had resulted in a lot of new objects being found; planets, asteroids, stars, nebula (many of which would soon be renamed galaxies), etc. Naming rights went to the discoverer, but the issue of who discovered what was often disputed and contentious. This was a time when the fastest international communication was limited to the speed of a train where possible, or more typically a horse, or in many cases, a boat. Cars were available, but rare, and roads were primitive where they existed at all.

To create a centralized clearing house for these names, and details for how to find these objects, and investigate the originality and veracity of new claims of discovery, the IAU was formed. Their original mission remained largely unchanged for most of their existence.

Previous to 2007, the IAU's mission statement (as posted on their website) was:

It's mission is to promote and safeguard the science of astronomy in all its aspects through international cooperation. [...] The IAU also serves as the internationally recognized authority for assigning designations to celestial bodies and any surface features on them.

— by means of http://web.archive.org

This is a grand and worthy mission. But the universe is an awful lot of real estate for one organization to have control over. Not only stars and galaxies, but also planets, asteroids, comets, moons, and every namable formation and feature on all of these celestial bodies.

After the whole fiasco around their definition of a planet, the language of the mission statement was changed. The new version says:

Its mission is to promote and safeguard the science of astronomy in all its aspects through international cooperation...

The key activity of the IAU is the organization of scientific meetings...

Among the other tasks of the IAU are the definition of fundamental astronomical and physical constants; ambiguous astronomical

nomenclature and informal discussions on the possibilities for future international large-scale facilities. Furthermore, the IAU serves as the internationally recognized authority for assigning designations to celestial bodies and surface features on them.

— https://www.iau.org/administration/about/ (Taken Aug. 27, 2021)

They retroactively added a sentence expanding their self-proclaimed authority to include "definitions ... of ambiguous astronomical nomenclature."

I have no problem with someone taking on the arduous task of organizing and cleaning-up the mish-mash of ad-hoc names and categories polluting the noosphere of astronomical nomenclature. But they've only continued and exacerbated the ad-hoc-edness. Their definition for a planet wasn't well considered. To claim the authority to do it after-the-fact was reactionary and defensive.

They also added, "The organization of scientific meetings," and "informal discussions on the possibilities for future international large-scale facilities." I think this is a clue to the real agenda behind the definition of a planet. Where money gets spent depends on what's important. What's important is affected by the labels we put on them.

Lisa R. Messeri writes that the IAU has given priority to one sub-group of astronomers over others. She describes three fundamentally distinct forms of cosmology:

Dynamic Cosmology concerns itself with movement and the interactions of gravitational fields—the study of orbits. These are the astronomers.

Structural Cosmology is concerned with what the celestial objects are made of—composition, geography and geology. These are the planetary geologists, or planetologists.

Cultural Cosmology is the third and largest group. This is the realm of the general public's historical perspective and common understanding, including most children, their parents, educators, and more importantly, voters and politicians who may (or may not) be inclined to approve funding for big projects and who are far more influenced by labels and impressions than actual scientists are.

— From: The Problem with Pluto: Conflicting Cosmologies and the Classification of Planets. Social Studies of Science, 40 no. 2, pp 187-214, www.jstor.org/stable.25677402

What the IAU did was not an act of "international cooperation," just the opposite, it was the voice of a small minority asserting dominance over the majority. It does not promote or safeguard the science. It ignores educators, historians, linguists, and more importantly, other groups of cosmologists, planetologist and physicists who are not specifically astronomers, and who adamantly promote the fact that Pluto and other objects are, in fact, planetary bodies.

Why didn't the larger group speak out at the time of the vote? Many of them have said they thought the issue of what is, or is not, called a planet just wasn't important.

> **Most planetary scientists call Pluto a planet and don't [care] what the astronomers say**
>
> — Alan Stern, in an interview with CBS News

If they had only been able to foresee how much controversy would result from that ill-fated vote, they might have taken it a little more seriously, and the outcome could have been different.

Would there have been a public outcry over Pluto retaining the status of planet? Would the "Pluto is not a planet" people have written so many papers and had so many interviews and written so many books about their position? Unlikely. Why? Because by any sensible definition of the word, Pluto is obviously a planet.

Another reason so many people may have abstained from the vote is because the IAU, like every other organization on Earth, has its own dynamics of internal politics and favoritism. These people have careers and professional relationships to protect. Brian Marsden, a long-time advocate for demoting Pluto to the status of 'minor planet', was also the head of the *Minor Planet Center* of the IAU from 1978 to 2006. If you wanted credit for your minor planet discovery, Brian Marsden was the gatekeeper.

> For a great review of the battle over who got credit for discovering the minor planet Haumea, please read the Wikipedia article at —
> **https://en.wikipedia.org/wiki/Controversy_over_the_discovery_of_Haumea**
> And also the alternative viewpoint at —
> **http://www.spaceobs.com/Blog-de-Alain-Maury/**
> **How-he-believed-he-killed-Pluto-alone-and-why-it-had-it-coming**

Who is Alain Maury? He discovered or co-discovered more than ninety (90) asteroids or comets in the Solar System. He has an asteroid named for him. He worked with Charles Kowal—one of the discoverers of the Centaur class of asteroids. In 2010 Maury proved that Eris was smaller than Pluto, re-establishing Pluto as the largest known object in the entire Trans-Neptunian region. Maury was co-author of the first paper to correctly determine an accurate diameter for Pluto, which was later confirmed by the New Horizons mission. In short, he has some credibility on the subject.

For the record, Mr. Maury agrees that Pluto should not be considered a planet (he *is* an astronomer, after all), but he strongly disagrees with the way it happened. Maury quit the astronomy profession because of all the in-fighting and puffed-up egos in the field. He said, "Seeing Sheldon Cooper is somehow fun on TV but dealing with similar nerds on a daily basis is tiring, and my life is too precious to waste it on and with idiots like that."

He goes on to say that many members are nice people, but they are drowned out by the greater majority. I especially admire his honesty about some of the personalities he has encountered. Regarding the rules for naming asteroids, he says, "The only problem is that Dr. [Mike] Brown never read them, or if he read them, decided to piss on them repetitively."

Alain Maury's blog entry about the whole "who killed Pluto" affair is a great read, especially about how astronomers who lack a PhD and are not from Caltech often get pushed aside and don't get credit where it's due. You can find the link under Alain Maury's name in the references section at the end of this book.

Alain Maury isn't the only one to accuse professional astronomy of bad politics. In the book, ***The Case For Pluto***, Alan Boyle writes:

If there's still someone out there who thinks that science and politics never mix, the story behind the battle of Prague should change your mind.

— Alan Boyle, The Case For Pluto, p. 114

The "battle of Prague" refers to the IAU meeting where the infamous vote on the definition for a planet occurred. The entire book is an amazing source of information about the personalities involved in the argument over Pluto and planethood in general.

Another great book on the subject is, ***Beyond Pluto***, by John Davies. On pages 186-189 he describes the touch-and-go battles for getting a Pluto fly-by mission funded. He mentions an influential stamp campaign showing detailed photos of all the planets except Pluto, which had the words "not yet explored" under an artists impression. It's worth noting that Davies' book was published in 2001, five years before the Prague vote. For an outstanding description of the politics involved in naming things at the IAU and who gets credit for discoveries, read the chapter, "Will we ever get our names right." in his book.

Mike Brown likes to take the credit for killing Pluto, but before the vote in Prague, he went on the record saying that he didn't care which way the vote went. His biggest contribution to Pluto's downgrade was the discovery of Eris. It was sheer luck that this object happened to have a larger mass than Pluto.

Both of the books mentioned above and Alain Maury's personal account point to two other astronomers who worked the hardest to exclude Kuiper Belt objects from the definition of a planet. Julio Fernandez was the biggest voice. He was also a key player in establishing the Kuiper Belt's existence. David Jewitt, the astronomer who discovered the first Kuiper Belt Object beyond Pluto said of him,

If anything, ... Fernandez most nearly deserves the credit for predicting the Kuiper belt

— www2.es.ucla.edu/~jewitt/kb/gerard.html

Fernandez was the leader of the opposition to the IAU's original definition during the vote in Prague. He and the people who stood with him forced the addition of the "clear the neighborhood" requirement.

They might not have gotten that far without the support and the authority of Brian Marsden. Marsden, the director of the IAU's Minor Planet Center from 1978 to 2006, had been working to get Pluto downgraded to minor planet status since the 1980s — way before the Kuiper Belt was discovered.

— *The Case for Pluto*, by Alan Boyle, p. 132

There is also an interesting timing in the push to downgrade Pluto. The New Horizons mission to study Pluto launched on January 19, 2006. At the very next general meeting of the IAU, only seven months after the launch of New Horizons, the IAU took its controversial vote that declared Pluto should no longer be considered a planet.

If the original proposed definition had been passed, it would have ended the controversy and been a pat on the back of the New Horizon's team. But for some astronomers, that just wasn't palatable.

There was no compelling reason why the IAU needed a definition at that time. Some people will claim that the discovery of Eris in January of 2005 (a year and a half before the vote) was what sparked the debate. But the debate over Pluto's planethood had been going on for more than 20 years by this point. The only repercussion the decision would have on Eris is whether or not it should be named according to the guidelines for planets, or according to a different set of guidelines used for minor planets. Would it be named minor planet 136199-Eris, or Planet Persephone? The IAU gives the discoverer *ten years* to make a decision, and there were more than eight years left to do it. This was not a pressing issue.

But it wasn't a fair fight from the start. No one had seen Pluto's clear image yet. There was no up-close data available, and there wouldn't be until New Horizons made its historic flyby. The structuralists had no real ammunition. Virtually everything we knew about Pluto was limited to what could be gleaned from its orbital parameters, and orbital parameters were all that the dynamicists cared about. This was the point they were making with that vote.

In the end, the IAU endorsed the sub-group of dynamicists at the expense of the structuralists and the public at large. They have generated confusion and dissent, as evidenced by the plethora of scientific papers, web pages, social media arguments and of course, this book and others.

Here's how they did it — In August of 2006 in the city of Prague, at the end of a week-long meeting and after the majority of participants had already left, fewer than half of the remaining attendees (a total of 424 people) representing only four percent of the total membership of the International Astronomical Union, took the fateful vote that would cause all the hoopla about Pluto and Planets. Many of them voted against the resolution, but a majority voted for it.

Amateur statisticians like to point out that a 4% sampling is more than adequate to gauge the majority opinion. But press a professional statistician on the issue, and they'll tell you this is only true for a genuinely *random* sampling of people. The people who voted Pluto out of the planet club were **not** a random sampling. They were a self-selected group with an agenda, not statistically valid for anything.

Shortly after, almost 400 planetary scientists who could not be present for the vote signed a petition against the IAU and their definition of a planet.

— https://www.nbcnews.com/id/wbna40470791

More people signed the petition *against* the definition than voted in favor if it. Most of the people who could have voted at Prague avoided it. When asked why, many said it didn't seem important enough to stick around for.

Why not? Because the vote was originally just a formality intended to confirm Eris as a planet, restore Ceres to planethood and do nothing to change Pluto's status. The announcement prior to the vote makes this perfectly clear:

"The world's astronomers, under the auspices of the International Astronomical Union (IAU), have concluded two years of work defining the difference between 'planets' and the smaller 'solar system bodies' such as comets and asteroids. If the definition is approved by the astronomers gathered 14-25 August 2006 at the IAU General Assembly in Prague, our Solar System will include 12 planets, with more to come"

-- https://www.iau.org/news/pressreleases/detail/iau0601/

No one expected the procedural ambush and last-minute changes that were forced into place at that vote. That's not science, and it's not how definitions work. It was nothing less than a political coup.

Chapter Thirteen: The Science of the IAU Definition

One of the foundational principles of science is that a hypothesis must survive being challenged before it can be confirmed. All you need is a single counterexample to invalidate a theory.

There are tests we can do to the IAU's definition of a planet that prove it to be inadequate as a scientific classification. This chapter will explore some of them. We'll see that the "clear the neighbourhood" clause was a hasty and ill-conceived last minute addition. Also, the final form of the wording for the whole definition is awkward and problematic. In short, it is a kludge of a definition, unsuitable for any scientific purpose whatsoever.

For the sake of reference, here it is again, copied word-for word in the original spelling and punctuation. AIU resolution B5, Definition of a Planet in the Solar System, reads as:

(1) A planet is a celestial body that

(a) is in orbit around the Sun,

(b) has sufficient mass for its self-gravity to overcome rigid body forces so that is assumes a hydrostatic equilibrium (nearly round) shape,

and

(c) has cleared the neighbourhood around its orbit.

Starting with the first statement we have a serious flaw. This is not a definition for a the *fundamental concept* of a planet. It specifically applies to the objects within our own solar system and no where else; "in orbit around the Sun," is not a mistake.

They did not intend to say "a star," and it cannot be construed to mean any star in any other solar system. The second paragraph of the IAU's resolution makes it clear:

The IAU therefore resolves that planets and other bodies, except satellites, *in our Solar System* be defined into three distinct categories in the following way... (emphasis added)

It is clearly applicable only to our Solar System.

A definition that is conditional and selectively applied is not a scientific definition for a fundamental concept. There's nothing fundamental about a single use-case.

Here's another problem with item (a). It's just a back-handed way of saying *No Moons Allowed.* Essentially everything in the Solar system orbits the Sun. There are millions of known things in orbit around the Sun, theoretically billions. Compare that number with only a few hundred known moons — the only things that don't technically orbit the Sun. But that's debatable since there's no technical definition of what an orbit is. (See the chapter on orbits for more about that.)

Also, since a moon is, by definition, contained within the orbit of a larger object, the "cleared the neighborhood" clause makes this rule completely unnecessary.

Now let's look at item (b). It turns out this awkwardly wordy clause is entirely superfluous too. Any planetary body large enough and massive enough to sweep its orbit clear will automatically be mostly spherical.

Depending on how you measure "round" there are between 110 and 184 mostly spherical bodies in the solar system. Of those, Pluto is the 17th *largest.* The addition of item (c) renders this part of the definition completely irrelevant.

So, the only real requirement for being a planet is to have a cleared neighborhood. Except, there's no guidance on what is meant by "clear" or "neighborhood." Why this oversight? Because item (c) was tacked-on at the last minute. The original definition for a planet had no such requirement. But during the vote, after a fierce episode of argument, item (c) was hastily added, specifically to exclude Pluto.

Item (c) is not especially useful for defining a planet. It is far more reliant on distance from the Sun and the age of a solar system than the size or quality of the object. The reason is simple. The farther from the Sun you go, the longer and slower the orbit will be and the longer it will take to clear it. Earth completes two hundred and forty-eight orbits for every single orbit that Pluto makes, and Pluto's orbit is 40 times longer than Earth's. In other words, it has a lot more work to do and fewer opportunities to do it.

The Kuiper Belt contains more volume than the entire Solar System inside the orbit of Neptune. An Earth-sized object deep in the Kuiper belt *could not* clear that orbit within the age of our Solar System. A Jupiter sized object in the position of Sedna *could not* clear that orbit. But we don't even need to go that far. If Mercury were in the position of Mars, it *could not* clear that orbit. We'll see the math to prove this later.

A clear neighborhood is not a defining quality for an object. It is a defining quality for a *situation*.

But it's also an impossible situation; there are no truly clear orbits in our solar system. Example: Earth. The *Aten-class* asteroids are a group more than 1800 near-Earth asteroids. *Interior-Earth* asteroids (also called *Atira* or *Apohele* asteroids) are contained entirely within the orbit of Earth. Asteroids that cross Earth's orbit include the *Amor-class* of asteroids (numbering more than 7400) and the *Apollo-class* of asteroids (the largest group, with more than 10,485). These are all officially recognized categories of asteroids within, across, and near Earth's orbit.

— https://www.spacereference.org Population numbers from Wikipedia.

Clearly, Earth's orbit is not "clear".

There's another class of objects called *Mars-crossing* asteroids. Recently asteroids have been found that cross Mercury's orbit too. Jupiter's orbit hosts hundreds of thousands of *Trojan asteroids*, and the *Centaur-class* of asteroids cross the orbits of all the gas giants and Pluto too.

Sure, you could apply one or more of the many suggestions floating around to refine the word 'clear'. But which one? And why that particular one? It's like all the definitions for a continent. The reader is left to interpret for themselves what is, or is not, significant. In other words, it's less of a hard definition and more of an *opinion*. (We'll do a deep-dive into this in the chapter on Clearing The Neighborhood.)

There are a lot of serious issues with the IAU's definition and how to interpret it. As mentioned, two of the three criteria for planethood rely heavily on the word orbit. Surprisingly, the IAU has no official definition for what an orbit is. But should they? Orbits seem obvious, right?

Think again. It turns out, there are a more types of orbits than most people would guess (more on this in the chapter on orbits).

Here is a little more nitpicking about the IAU's definition. (Note: these are not the product of my solitary imagination. These were the more common complaints gleaned from various on-line forums.)

If Jupiter were replaced with a primordial black hole of the same mass, it's a planet. In fact, you can replace all the planets with equal mass primordial black holes and all the orbits of everything in the Solar System will remain exactly the same. Since *primordial* black holes were never stars, and are also more spherical than any of our current planets, they'd fit the definition for planets too.

While we're on the subject of roundness, the actual words are "hydrostatic equilibrium". Hydrostatics is a study of fluids at rest. The prefix "hydro" generally means water, from the ancient Greek word for water. Hydro*dynamics* is the study of fluids in motion.

Once a planet is squeezed into a sphere by its own gravity, it's no longer at rest. Stuff get pushed around under the pressure. Things start to move and the planet becomes geologically "alive." It is no longer hydro*static*. It has become dynamic.

More importantly, this kind of analysis is not the purview of astronomers and their telescopes. This is the realm of planetary geologists—the vast majority of whom consider Pluto to be a bonafide planetary body.

Round is a profoundly bad metric. Too much spin can deform a spherical object, like Saturn. Its oblateness is easily visible to the naked eye. Saturn is only round when measured around its equator. It is not round when measured from pole to pole. It is 10% shorter than it is wide.

Different materials such as ice, rock, and iron all achieve a gravity-induced spherical shape at vastly different masses. A "planet" of ice can have far less mass and less gravity than a large bumpy, lumpy metallic asteroid. Consider a hailstone, or a raindrop, as compared to the Eiffel Tower. One is more round, the other is more massive.

The tiny ice-moon Miranda is a satellite of Uranus with a mass of 6.4×10^{19} kg. It is more spherical than the much larger Pallas, the second largest asteroid with a mass of 20×10^{19} kg. — triple the mass of Miranda!

Miranda measures 480 x 468 x 465.8 km. It is out-of-round by only 3%. Pallas is out-of-round by 16.5%. But Haumea has a mass of 200×10^{19} kg. It is ten times the mass of Pallas and more than 30 times the mass of Miranda. It is considered a legitimate dwarf planet, officially sanctioned by the IAU. Therefore it must have an acceptable value for "roundness". But it measures 2100 x 1680 x 1074 km. It is out-of-round by 200%!

I'm sorry, but that isn't round. Even the IAU ignores their own definition. It's that bad. Roundness is not a reliably consistent criteria, and it's a pointless to have a requirement that we're just going to ignore anyway.

Let's move on to orbits.

Jupiter does not actually orbit the Sun. Hard to believe, but true. Jupiter's mass is so great that the center of mass in the Sun/Jupiter equation (the barycenter) is actually 30,000 miles *above* the surface of the Sun. Both Jupiter and the Sun co-orbit an empty point in space between them. Thus, Jupiter is not a planet.

Jupiter's path around that barycenter is also far more polluted with objects than Pluto's. There are credible estimates that more than a million asteroids could be locked in the same orbit as Jupiter. They are collectively known as the Trojan, Greek and Hilda asteroids.

Some people want to believe there's a specific ratio for the mass of the object and the mass of the clutter. Other people think objects at Lagrangian positions of an orbit shouldn't count. But that's not part of the official definition. It's an ad-hoc ex-post-facto modification.

When your reasons for believing something are justified ad-hoc, you are left susceptible to further discoveries undermining the rationale for that belief.

— Neil deGrasse Tyson, pg. 51, The Pluto Files.

We already know of exo-planetary examples that defy these ad-hoc modifications. There is credible scientific evidence for a planet several times the Earth's mass lurking deep within the Kuiper belt. If (when) this object is found, it will not be a planet, but a dwarf planet as much as six times the mass of the Earth.

It is noteworthy that the exception made for Haumea's profound lack of roundness is based on the fact that it spins too fast to be round. But no exception is made for Ceres's orbit not being clear due to the external effects of Jupiter's influence. It seems that the different parts of the definition are being somewhat selectively applied.

That ain't science.

Chapter Fourteen: An Exoplanetary Example

Twenty-seven percent (27%) of the star systems within 100 parsecs of Earth are multiple-star systems. This includes binary stars, triple stars, quadruple and even quintuple star systems. There are also systems with no stars. A lot of zero-star systems exist within the spaces between the stars. These zero-star systems are centered around free-floating "rogue" planets, usually the size of Jupiter, that harbor a fleet of planet-like satellites just like Jupiter does. If we intend to define things by their orbits and what's at the center of their orbit, we're gonna need a lot more definitions and sub-classes within those definitions.

Let's do a thought experiment: Two stars are in orbit around their common center of gravity. One star is about 25% larger than the other. The distance between them is about 20 AU, roughly the distance from the Sun to Uranus. At that distance, their two habitable zones do not overlap. There is plenty of room between them for two systems of rocky planets.

It would be easy to think of these rocky objects as planets; they do orbit a star. But the lesser star orbits the larger one in the same sense that Charon orbits Pluto, or Jupiter orbits the Sun. Both stars actually orbit a point in empty space between the two of them.

Since two of the moons that orbit Jupiter are larger than the planet Mercury, but are still considered moons, does this mean the planets around the lesser star in our two-star system should be classified as moons too?

Remember, the IAU's definition of a planet is more about orbits than the object. The only requirement for the object is to be round. Also, Jupiter is more like a star than like a planet. It emits more heat and radiation than it receives, is highly radioactive, has no solid surface…

If Jupiter were a small star, (say, a minimally sized red dwarf) would Io, Europa, Ganymede, and Callisto be called planets? What if Jupiter were a brown dwarf — not a planet, but also not a star. Do the moons of Jupiter become planets then? If this seems absurd, then the idea of defining objects by whatever happens to exist at the center of their orbits should also seem absurd.

Let's continue—

Between the Sun and Jupiter is point in space called a Lagrangian position, also known as L1. Unlike the barycenter (which is closer to the high-mass object), the L1 position is closer to the lower-mass object. The reason it is called L1 is because there is also an L2, L3, L4 and an L5 at specific locations in the space around two orbiting objects.

The Lagrange Points

In the diagram below, the Sun is at the center, the orbiting body is between L1 and L2. All of the L<u>x</u> points are empty space.

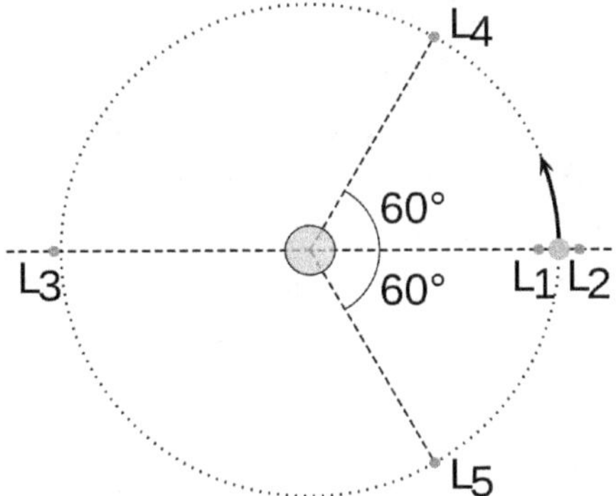

The positions L1, L2 and L3 are like being at the top of a hill. As long as you don't drift very far you can sit there without any effort. But move too much and you'll roll down the hill. It's a great place to park a spacecraft that can make minor adjustments to stay at the top of the hill, but any natural objects placed there will eventually fall out.

L4 and L5 are unique. They are gravitationally stable. Being in one of these positions is like being at the bottom of a bowl. A small object can orbit the empty space around the L4 or L5 positions like swirling round in the bowl. This is actually common in the Solar System. In the case of the Jupiter/Sun L4 and L5 positions, there are millions of trojan asteroids orbiting these empty points in space. By some estimates, there may be more asteroids swarming in Jupiter's L4 and L5 points than in the asteroid belt.

In the case of our theoretical two-star system, there could be objects in orbit around their mutual L4 and L5 positions too. Since these objects are gravitationally bound to both stars, but don't orbit either star, are they planets?

This is actually an unlikely scenario, but not impossible. It's unlikely because the mass ratios are not favorable in this example. But it's not impossible either—two of Saturn's moons have asteroids orbiting their L4 and L5 positions, an equally unlikely scenario.

Now, suppose this dual-star system also has a large planet in a far distant orbit around the system's barycenter. It is gravitationally bound to both stars but is not orbiting either of them; it orbits the common center of gravity between them. This planet is about eight times the mass of the Earth and it about 2,000 Astronomical Units (AU) from the barycenter of this system. At that distance, the orbit is too slow, and the volume of space is too large for it to be cleared. It's impossible for this super-Earth to "clear its orbit". It is technically a dwarf planet, eight times the size of the Earth.

Let's make it worse. Increase its distance from the dual-star barycenter to more than 8,000 AU, and increase the mass of this dwarf planet to ten times the mass of Jupiter. It's a super-Jupiter. It's still not a planet since it will still have a lot of co-orbital debris out there. But it does have moons, and one of these moons is as large as the Earth. We now have a giant dwarf planet with an Earth-sized moon.

Changing nothing else, let's increase the mass of the super-Jupiter planet to 50 times the mass of Jupiter. This will change it to a brown dwarf—still not a star, but also not a planet. What's the status of its Earth-sized moon now?

Let's increase the mass of the brown dwarf again, this time to 125 Jupiters. Now it becomes a red dwarf—a real star. The Earth-sized "moon" hasn't changed, but now it could be called a planet, I suppose? Is an object that orbits a red dwarf that orbits the center of gravity of two other stars a planet? Or a moon? Or something else? Keep in mind there could also be planets orbiting both of the main stars and perhaps even a few planet-sized objects orbiting the dual-star L4 and L5 positions.

Since our definition for a planet has no provisions for these kinds of orbits, we must rely on our common sense instead. We must ignore the orbits and classify the objects according to their inherent properties.

Could a system like this actually exist? In fact, it *does exist.* And it is very well known. This scenario describes our closest stellar neighbor, the Alpha Centauri system.

The Alpha Centauri system is a triple-star system. The largest of the three stars has the official names of Rigil Kentaurus, also known as Alpha Centari A. It is 1.1 times the mass of our Sun. The second star is Toliman, also known as Alpha Centauri B, at 0.9 times the mass of our Sun. Lastly is Proxima Centauri, the smallest star, at 0.125 times the mass of the Sun.

Rigil Kentaurus and Toliman orbit each other at approximately the distance from our Sun to Uranus (It actually varies from the distance of Saturn's orbit to the distance of Neptune's orbit). Proxima Centauri orbits their common barycenter about 430 times that distance (about 8000 AU), and it is known to have an Earth-sized natural satellite orbiting it.

So far, no planets have been discovered closer to the center of that system. It's virtually impossible to distinguish an object parked between two stars from four light-years away. But people have modeled the evolution of this system on computers—a lot of people have. Regarding the two inner stars:

In each instance, despite different parameters, multiple terrestrial planets formed around the star. In every case, at least one planet turned up similar in size to the Earth, and in many cases these planets fell within the star's habitable zone.

— UC researcher Javiera Guedes, in an interview with Wired Magazine.

Other scientists, using different methodologies, estimate an 85% chance that planet-sized objects exist in the habitable zone of Alpha Centauri's two main stars.

— https://vimeo.com/174313049

Connecting these dots to our own system — Charon is considered a moon of Pluto. Jupiter is considered a planet of the Sun. In each of these cases, the larger and smaller objects both orbit a common barycenter that is external to either object. In both cases, the lesser sized object gets the diminutive classification.

What does this mean for the lesser stars of the Alpha Centauri system and the objects orbiting them, and the objects orbiting those objects? Should we work to create specific definitions for every potential combination of star, brown dwarf, gas giant, rocky planet, asteroid and moon scenarios in one, two, three, four, five, and zero star systems? This would probably be a fruitless exercise.

All these new planets are incredibly diverse. Nature is more creative than most of the imaginative scientists out there.
— Jack Lissauer, co-investigator, Kepler Mission.

It is simply not necessary. Defining an object according to the quality of whatever is at the center of its orbit and the other objects around it is far more complicated than defining an object based on its intrinsic properties, regardless of where it is or what's nearby.

Alan Stern pointed out that a mountain doesn't stop being a mountain just because it is surrounded by other mountains. Unlike the IAU's concept of a planet, most other types of objects in astronomy are based on *what* they are, not *where* they are or what else might be near them.
— https://www.nbcnews.com/id/wbna40470791

The orbit doesn't need to be included in the classification scheme for planets. Gravitational dynamics for orbiting masses is already well understood as a separate thing, independent of the type of objects involved. The only things the dynamicists need to know are the dimensions of the orbits and the masses involved.

Perhaps the IAU can create a list of basic orbital types for the more common situations, but it must also acknowledge that this defines orbits, not objects. As we'll see in later chapters, any kind of object can be in any kind of orbit.

Chapter Fifteen: Clearing the Neighborhood

Since 'clearing the neighborhood' is the most problematic part of the definition, I've dedicated a whole chapter to it. This part of IAU's definition is so deeply flawed it's actually offensive to a lot of scientists and logic loving lay people like myself. It does not work. There is no explanation given and no agreement on what is meant by "clear". There is also no explanation given and no agreement on how big a "neighbourhood" is.

According to Steven Soter, there is more volume in the Kuiper Belt than in all the space between the Sun and the orbit of Neptune.

— "What is a Planet?" *Astronomical Journal 132*, Dec. 2006

We know there's a lot of 'stuff' inside Neptune's orbit—you and me and our home planet included. So, what are the limits of a neighborhood? Can any definition be called scientific that is so dependent on such vague and imprecise terminology, wide-open to interpretation? And if it's not a scientific definition, and it's not aligned with the common and historical use of the word it pretends to define, then what use is it at all?

Unfortunately, now that this steaming pile of word-stuff has been established and placed on a pedestal with spotlights, it can't just be ignored. Teachers and kids will still want to know what a planet is. Can we fix the definition for them? Some scientists have tried. But, you know, scientists…

Rather than giving the kids a cumbersome list of a dozen or even two-dozen planets, Jean-Luc Margot has helpfully suggested clearing up the definition with a simple equation. He says,

This relationship clearly distinguishes the 8 planets in the solar system from all other bodies.

— https://arxiv.org/abs/1507.06300

Here's that equation:

$$\frac{M_p}{M_\oplus} \gtrsim C^{3/2} \left(\frac{M_\star}{M_\odot}\right)^{5/8} \left(\frac{t_\star}{1.1 \times 10^5\ \mathrm{y}}\right)^{-3/4} \left(\frac{a_p}{1\ \mathrm{au}}\right)^{9/8}$$

Hooray! Now we have a technical solution that does what the IAU wanted, and it works for exoplanets too. It's beautiful! Look at those exponents! Elementary school teachers rejoice!

I'm kidding, of course.

This equation for defining a planet reminds me of a classic math exercise: "What's the next number in this sequence: 1, 3, 5, 7, __ ?"

If you said 9, sorry, that's wrong. The correct answer is 217314.

Here's how:

f(x) = (18111/2)x^4 - 90555x^3 + (633885/2)x^2 - 452773x + 217331

f(1) = 1

f(2) = 3

f(3) = 5

f(4) = 7

f(5) = 217341

Simple, right? (Thanks to Avinash Sahu for this equation!)

If we favor simpler definitions, then Mr. Margot's equation for a planet utterly fails in that regard. Apparently someone else noticed that too. Here's a simplified equation for Margot's concept:

$$\mathbf{m / (M^{5/2} * a^{9/8})}$$

m = *the mass of the planet,*

M = *the mass of the parent star,*

a = *the distance from the star to the planet.*

Margot wasn't the only one with a method to quantify clearing the neighborhood. Astronomers Stern and Levinson proposed a value called lambda (Λ). If the lambda value is greater than 1.0 for an object, that means it has the ability to clear its neighborhood within the lifetime of the universe, so it must be a planet.

Here's their equation:

$$\Lambda = m^2 / ((a^{3/2}) * k)$$

m *= mass of planet,*
a *= orbital distance*
k *is a fudge factor.*

The advantage of this one is that you don't need to know the mass of the parent star. But you do need to calculate a different fudge factor for each solar system, and finding that fudge factor probably does depend on the mass of the star.

Astrophysicist Steven Soter offers the simplest equation for a planetary determinant. His is simply the mass of the planet (M) divided by the total combined mass of anything that crosses its orbit even once.

$$\textbf{planetary determinant} = M / \sum m$$

Simple, right? Now all we need to do is survey the entire solar system, calculate the trajectories of everything, project their paths into the past and also into the future, then sum up the mass of the items that have crossed, or will cross the orbit of our candidate planet. Then we can finally know if we've got a genuine planet or not.

Since all the proposed categorizations are essentially arbitrary filters, we may as well define a planet as, "Whatever the IAU damn well pleases." Because that's basically what's happening, but with extra steps. Meanwhile, kids are being alienated and discouraged from studying astronomy by all this nonsensical brouhaha.

Someone is likely to point out that all of these equations come to the same conclusion, that Pluto stands apart from the rest of the planets. And they are right, because that's exactly what all of these equations were specifically designed to do from the start. Like the sequence 1, 3, 5, 7, ___, we can put whatever value we want in that blank space and find an equation to satisfy this progression.

Here's a popular graph showing a convenient line of demarcation:

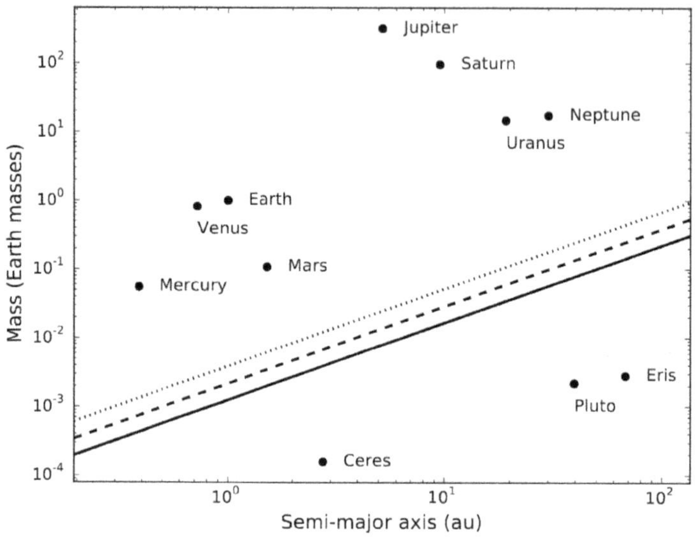

— From: A Quantitative Criterion For Defining Planets, Jean Luc Margot

The three lines represent different distances from the object (measured as a ratio of its gravitational influence, known as its 'Hill Sphere') and different amounts of time for the clearing process to take place, measured in billions of years. In younger solar systems the lines will shift higher. They will be more chaotic, and therefore will have fewer "planets" due to insufficient time to "clear the neighborhood". All planets in those young solar systems could be called dwarf planets according to the IAU's theory, regardless of how large they are.

This graph also illustrates one of the biggest problems with the "clear the neighborhood" requirement. The farther out you go from the Sun, the larger an object has to be. If we increase Eris's mass to equal Mars, it would be a dwarf planet larger than the planet Mercury. Other astronomers point out that even Earth would not be able to 'clear the neighborhood' at the distance of Eris. (They use different math than Margot does to reach their conclusions, which is perfectly valid since there is no official definition for 'clear the neighborhood'.)

And, as Philip Metzger points out, this specific chart only works as intended for *our* solar system. The lines will also shift up or down depending on the mass of the star. Within the habitable zone of a red dwarf star (the most common type of star in the galaxy), an object the size of Saturn's tiny moon Tethys would qualify as a planet. But in the habitable zone of a larger O-type star, the Earth is two orders of magnitude too small to be a planet.

Metzger created this graphic to illustrate the problem.
This is Tethys and Earth shown at the same scale.

Here is yet another mathematically-based determinant for a planet:

$$\mathbf{2 <= M/(9x10^{20}) + D/800}$$

M = mass of the candidate planet in kg,
D = its diameter in meters.
($9x10^{20}$) is the minimum mass for a "planet"
800 is the minimum diameter for a "planet".

Anything with a value greater than two is considered a planet. In this case, Pluto and Ceres and Eris are all confirmed as planets. None of the other asteroids and very few of the KBOs will qualify.

If you want to exclude Ceres, Eris and Pluto, just change the minimum values. You can choose minimum values for whatever dividing line you want.

But does any of this advance the science? Does it inspire anyone to respect the science of astronomy and yearn to follow that path? Or is it just another arbitrary line that serves no real purpose other than to have a line.

There is an old idiom: If it looks like a duck, walks like a duck and quacks like a duck, it's a duck. That may not satisfy a rigorous analysis of the genome, haplotype and epigenetics of similar-seeming duck-like organisms (or crab-like organisms), but it teaches a simple truth to the kids as they continue their education. Science is not about the decree of certain authorities, or voting, or most certainly not about cherry-picking specific details to fit a predetermined outcome. It's about observing facts, all the facts, and discovering patterns, causes, and effects based on them. Then we refine our understanding as we gather more information. It's a process, not a dogma. As kids learn, the more they learn, the more refined their understanding can be. This is education in action. The more education we attain, the more precise we can be in our nomenclature.

Planet is not a PhD-level word. It's a common word, for common people, like the word *crab*. No serious biologist would use the word 'crab' as the only descriptor for her research. She'd use the scientific name that carries the most specificity to adequately describe the organism she's studying and writing about. But scientific research papers and scientific terminology are are used to communicate with other scientists doing science, not for kids or the general public. Those vocabularies are not the same. No one says "Let's celebrate with Paralithodes Camtschaticus for dinner." But even tenured professors of marine biology will say, "Let's have crab legs for dinner."

The IAU is trying to apply a scientific rigor to something that is, and always has been, a common word. To say Pluto and Ceres are different *types* of planets is a useful application of the word for use in casual conversation among the whole population, regardless of a person's age, maturity or education level attained.

One thing is certain—we can never go back to a nine-planet solar system. Calling Pluto the last planet is misleading. It obscures the fact that another pluto-sized object exists out there, specifically, Eris. Leaving Pluto and Eris off the roster of planets reduces the likelihood that kids will learn about them, or other things like them.

Like it or not, people have been taught to expect a discrete number of planets in the system. This is misleading. It gives the impression that there is a definite natural boundary. But there is no boundary between major planet and "dwarf" planet in nature.

So, how should we answer the question of how many planets are there? As we'll see in the next chapter, the only honest answer is, "how many planets do you want?"

Chapter Sixteen:
How Many Planets Do You Want?

Let's put the argument that "no one wants hundreds of planets in the solar system" to rest. No one's head is spinning from too many planets in the Star Wars and Star Trek universes. Why not have a *real* universe of planets to explore?

If you want to build a ship, don't drum-up the men to gather wood, divide the work, and give orders. Instead, teach them to yearn for the vast and open sea.
— Antoine de St. Exupery.

Kids love to learn and explore. If you want to spark a kid's imagination tell them there are unexplored planets out there just waiting for someone to study them. If you want to squash a kid's imagination, tell him or her there are only eight planets, that's all there will ever be, and they were all found and cataloged a long time ago. That's not a trend anyone will want to follow. It's a great way to diminish the grandeur of the Solar System to the status of old, dumb and boring.

There are twenty-six letters of the alphabet that no one seems to have trouble with. We have fifty states. We seem to do fine with hundreds of countries on Earth. No one's making a fuss that we have too many countries, and could we please demote the landlocked ones to sub-country status? There are 88 official constellations in the sky. There are 94 naturally occurring elements.

In 1995 Pokemon originally started with 150 unique characters and has continued to expand ever since. Today (as of Sept., 2021) there are 896 characters in the Pokemon universe. And yet, it has only *gained* in popularity. Kids flock to it and absorb all the details they can about as many

Pokemon as they can collect. But somehow, thirteen, or thirty-two, or a hundred and ten planets is too many?

In a 2010 interview with NBC News, Alan Stern said, "The IAU definition was specifically designed to limit the number of planets to a small number. That, in my opinion, is not scientific. It's somewhere between personal preference and political."

In a 2006 interview with NPR, Mike Brown, the self-proclaimed Pluto Killer said, "We can't have 50 planets. My little daughter won't be able to remember all their names."

In the words of Alan stern, "This is hardly a scientific rationale."

Meanwhile, my own daughter's circle of middle-school friends memorized all the bones in the human body, then moved on to see how many digits of pi they could reliably memorize. *Reliably* was important to them. Accuracy mattered. My daughter got up to 40 digits; it was the passcode on her phone for a couple of months.

Mike Brown has also said, "Mercury, Venus, Earth, Mars, Jupiter, Saturn, Uranus and Neptune — are clearly in a class by themselves, a class Pluto, Eris and similar bodies just clearly do not belong to."

But hold on a minute. Do the gas giants really belong in the same class of objects as Mercury and Mars? It turns out that Isaac Asimov and Carl Sagan already had that conversation. Afterwards, Carl Sagan reported:

The solar system consists of the Jovian planets—Jupiter, Saturn, Uranus, Neptune—and debris. We are part of the debris.

— Rolling Stone Magazine, Carl Sagan: Life On Other Planets?, A conversation with the astronomer about the Mars Mariner Project, June 7, 1973.

So, there it is. We have it on good authority, predating the IAU's move by more than three decades, and by two of the top names in the field, that there are only four planets in our solar system and some debris. The gas giants are clearly in a class by themselves. They are profoundly different from the rocky planets.

It also seems pretty clear that the Earth and the other rocky planets are more than just debris. Also, Ceres, Pluto, and Eris are profoundly different from the asteroids and comets they swim with in the cosmos.

With all due respect to Mr. Sagan and Mr. Asimov, we need a home that's more than just "debris", and we need vast and open frontiers upon which to cast our dreams of exploration and discovery. We need more planets in this solar system of ours than just four, or eight.

Fortunately, one pair of venerated scientists do not have the authority to decree what the word 'planet' shall mean for the rest of us. How about seven? How about four hundred scientists casting a vote on the issue?

That's not how definitions work, and it's definitely not how science works. With four planets firmly established as a minimum, let's continue to explore the idea-space and see what else we can learn.

If you look at the Solar System the way Mike Brown does, then he's right. The classical planets' *orbits* are clearly in a class by themselves—except for Mercury. Its orbit is almost as eccentric as Pluto's. But Mike Brown has said he's willing to give up Mercury too. (What's with this guy?)

If we abandon Mercury, the *seven* planets' orbits are more round and distinct and different from the orbit of Pluto, and pretty much everything else skittering about beyond Neptune. The seven planets' orbits seem to be truly unique.

Well, except for most of the asteroids in the Asteroid Belt, that is. The asteroids' orbits are (for the most part) just as regular, just as flat to the ecliptic and follow all the same general trajectories as the seven planets, excluding Mercury because of its wildly eccentric orbit.

When you ignore the properties of these objects and look at nothing but the quality of their orbits, the categories are very different than if you look at the objects themselves.

This is what the IAU tried, and failed, to reconcile. Their definition is a verbal venn diagram declaring that "planets" must sit in the intersection of three sets: A planetary orbit, a planetary body, and a private neighborhood.

Unfortunately, the intersecting region is not consistently reliable.

Ceres doesn't fit the definition, but a Ceres-sized object in orbit between Saturn and Uranus would. If you put a primordial black hole of the same mass as the Earth way out where Eris is located in the Kuiper Belt, it would not be able to clear that orbit.

Some readers may be thinking that a black hole would surely "suck-up" everything out there. But that's not actually how they work. A 'primordial black hole' is another name for a dwarf black hole that's been around since the beginning of the Universe. Gravity for these objects works exactly like the gravity of any planet, or gas giant, or brown dwarf or small star. Its influence on its neighbors depends only on its mass.

Primordial (dwarf) black holes are theoretically out there, and could be as common as stars. They could even be in orbit around stars. If you think this idea is crazy, don't blame me. I didn't make it up.

There is solid evidence for an object deep in the Kuiper belt with a mass of five or six Earths, but no one has been able to find it. One of the theories for why we can't find it is that it might be a baseball-sized primordial black hole. The evidence for this object is the same mathematical reasoning behind the discovery of Neptune. There is a "weirdness" to the orbits of the largest objects in the Kuiper Belt that is best resolved by a large mass in a large orbit. Ironically, Mike Brown was one of the astronomers to work this out, but he rejects the idea that it might be a primordial black hole. Any object, even one bigger than the Earth, could just be too dark to see at that distance.

(*An abundance of Primordial black holes is one of the proposed solutions to the problem of what dark matter is. But we haven't found any yet, so we can't say for sure if they actually exist. It's hard to see something that's as dark as dark can get, especially when it's surrounded by the black of space and it's the size of a grapefruit, and it is millions of billions of miles away, or farther. But our current scientific models of the universe say they should be there.*)

If primordial black holes, and billionaire's sports cars, and brown dwarfs can all be in planet-like orbits while not being planets, can planet-like objects be in non-planetary orbits? And if so, what should we call them?

In 1610 Galileo was first to use a telescope to look at the celestial objects. But other astronomers immediately followed. The German astronomer Simon Marius discovered the four moons of Jupiter at approximately the same time as Galileo, but Galileo was more famous, more popular, and was given the credit. Which of the two men was actually first is still disputed by astro-historians.

Marius wanted the four newly discovered satellites of Jupiter to be called the Saturn of Jupiter, Jupiter of Jupiter, Venus of Jupiter, and Mercury of Jupiter. The names didn't stick, but the four Jovian moons were clearly considered comparable to another solar system.

Saturn's moon Titan was discovered next, in 1655, by Christiaan Huygens. He called it Saturni Luna, or "Moon of Saturn". Huygens started the trend of referring to satellites as moons. Some habits are hard to break.

Titan was known as the largest moon in our Solar System until fairly recently. If Huygens had known Titan was larger than the planet Mercury, we have to wonder what other precedent he might have started. Would he have followed Marius's ideas and called it a planet instead?

While we're on the subject of gas giants and their moons—Titan is no longer the largest moon in the solar system. Now Jupiter's moon Ganymede takes that title. What happened?

Titan has a thick, dense, opaque atmosphere. Early astronomers measured Titan's diameter across its visible gas envelope. Once we learned how thick that envelope is, and subtracted it from the diameter of Titan's visible 'disk', it no longer qualified as the largest moon. Ganymede took the title.

Take note: An atmosphere is a gas envelope. It doesn't count toward a moon's diameter. It also doesn't count toward a planet's diameter. Earth and Venus and Mars are planets who's diameters are measured at the bottom of their atmospheres, not the top.

Now, considering this, how are we going to measure the gas giants? They are nothing but atmosphere that gets gradually denser and denser, until it's dense enough that we could call it a liquid, but there's no discrete boundary there. It's not like sea level on Earth. You can't float a boat on this liquid, because by the time you reach a level where it's dense enough to float a boat, you're already sunk. Lower still, this liquified atmosphere gradually transitions into a slushy sloshy 'core.' We have no idea how far down this goes until something we can call solid finally comes into play.

According to an article on Space.com by Tereza Pultarova, Saturn's core is "fuzzy." According to Matt Hedman, Planetary Scientist at the University of Idaho, "Saturn's core is extremely diffuse." They estimate the core of Saturn may be 60% of the width we typically apply to the ringed gas-ball.

Since atmospheres don't count toward the diameter of a planet, and there's no discernible "sea level" to work with, then the reported sizes of all the gas giants *are not valid* if we insist on calling them planets.

Someone is sure to point out that we can explore them with balloons or some kinds of hovering craft. Sure. We can do that on Earth, Venus and Titan too. As I write this, we have a hovering drone working on Mars.

Atmospheres still don't count towards a planet's dimensions. And if we're going to create special exceptions whenever it's convenient, why stop there? Are we doing science or just playing favorites?

If the categorical status of planetary objects is so vitally important, why has no one addressed this before? Since they have no surface nor sea level, can the gas giants be called planets at all?

Titan is the most planet-like object in the Solar System after Earth. It is the only other object in the Solar System with lakes and rivers. It has clouds and rainfall too, but not of water. Titan is way too cold for that. Water on Titan is like lava, and ice is a kind of rock. It rains liquid methane. Earth's atmosphere has methane too, but not as much, and it's always in gaseous form due to our higher temperature.

Titan and Earth are the only planets in the Solar System with a nitrogen-rich, dense atmosphere. Titan's atmosphere is even thicker than Earth's. Titan also has seasons. Its surface is young, indicating a robust geology that reshapes the surface, just like Earth. And it is larger than the planet Mercury. Titan is **by far** the most planet-like object in the Solar system by comparison to the features and systems of Earth. It just happens to be the largest natural satellite of Saturn.

Titan, and the four largest natural satellites of Jupiter: Io, Ganymede, Europa, and Callisto, all have more in common with the rocky planets than any of the gas giants do.

If we remove the four gas giants from the category of "planets" and add their largest planet-like natural satellites, we'll gain seven new planets for our list: Io, Ganymede, Europa, Callisto, Titan, Triton and Titania. Those last two are the seventh and eighth largest natural satellites in the Solar System. Our moon is the fifth largest natural satellite.

So now our tally is... Wait, I lost track. Are we still including Mercury or not? Did we make a decision on Earth's moon? What about Ceres? Also, Triton has a retrograde orbit — it's the only large object in the Solar System to do that. Can we allow that or is it too "weird"?

Let's back-track a little and review the process of discovery for some of these things, and how it affected the planet count throughout history.

Venus was the first planet discovered, around 1600 BCE by the Babylonians. A few centuries later they discovered four more: Mercury, Mars, Jupiter and Saturn. Uranus was discovered in 1781 by William Herschel. Uranus is actually visible to the naked eye under ideal conditions.

It may have been catalogued as far back as 128 BC and included in a star catalog by an ancient Greek astronomer named Hipparchos, but it moves so slowly it was thought to be a star. Herschel gets the credit for Uranus's discovery because he was the first to notice its motion.

Uranus was originally classified as a comet. But observations by other astronomers convinced everyone that it was a planet. This was exciting for two reasons; of course finding a new thing is always a thrill, but finding Uranus seemed to confirm a theory that eventually helped in finding Ceres.

In 1772, long before Le Verrier did his celestial mathematical wizardry, a man named Johann Elert Bode predicted that a planet should exist between Mars and Jupiter. Working with astronomer Johann Daniel Titius, they created the Titius-Bode law. It's a purely mathematical progression that seems to correlate to the distances between the Sun and each of the then-known planets. In a nutshell, each planet is roughly twice as far from the Sun as the previous planet. Except, according to Bode's law, there was a missing planet between Jupiter and Mars. Either Bode's Law was wrong, or another planet was waiting to be found.

Before that planet could be discovered, Uranus was found way out in the far reaches of the Solar System. When an accurate distance was calculated, it fit the pattern of Bode's law. This confirmed it, and the first gold-rush style hunt was begun for the missing planet between Mars and Jupiter. (I hope you're getting the point here — planets are exciting! When was the last time anyone got excited about a new asteroid discovery?) Before long, Ceres was found right where it should be, or as they say, "close enough for government work."

That was in 1801. Ceres was immediately crowned as the new eighth planet. Soon after, four more objects in a similar orbit were discovered and dutifully added to the list of planets. Welcome to the twelve-planet Solar System!

Then in 1846 Neptune was discovered by Le Verrier and officially became the 13th planet. But Neptune's orbit doesn't fit the Titius-Bode law. This was a big problem. Science—*real science*—doesn't tolerate exceptions to fundamental concepts. Later, further refinements of all the planets' orbits showed that the Titius-Bode predictions are only loosely approximate. It's a pure coincidence that it fits at all. The Titus-Bode "law" was summarily abandoned.

Even though Bode's law wasn't accurate, it helped to inspire people to look for new things, and new things were found.

More Ceres-like objects continued to be discovered, and by 1850 the list of planets numbered as high as 23 or 32, depending on who's list you referenced (some included the satellites of gas giants and a few comets.)

But there were two new problems. The first problem was with resolution. The planets we now call asteroids were too small to resolve into a "disk" when viewed through the telescopes of the time. (It still bugs me that astronomers call a sphere a *disk* when viewed through a telescope. Astronomers...) No matter how much they zoomed in, the asteroids remained points of light, just like the distant stars. In all fairness, if Mars or Mercury were as distant as Uranus or Neptune, they would have the same problem.

The other problem was naming all these new planets. This could be a contentious process. Not only did new objects need a name, but the conventions of the time required them to have a symbol too, and the symbols were getting complicated. Here are the nineteen official planetary symbols of 1850 alongside their German names. (The Sun and Moon were still included by tradition.)

Bezeichnung
der Himmelskörper.

☉	**Sonne.**	♃	**Jupiter.**
☾	**Mond.**	♄	**Saturn.**
☿	**Merkur.**	⛢	**Uranus.**
♀	**Venus.**	♆	**Neptun.**
♁	**Erde.**	⯙	**Asträa.**
♂	**Mars.**	⯚	**Hebe.**
⚶	**Vesta.**	⯛	**Iris.**
⚵	**Juno.**	⯜	**Flora.**
⚴	**Pallas.**	⯝	**Metis.**
⚳	**Ceres.**		

In 1854, Johann Franz Encke promoted the idea of using a simple number in a circle for all future symbols. It was an immediately popular idea among astronomers. (Anything that makes a job easier will be popular.) But it also created confusion. Did the numbers represent their order in distance from the Sun, or the order of discovery? Those two numbers were not the same. In practice the number represented the order of discovery, regardless of where the object was in relation to the others. But not every publication understood this. It got messy.

Between 1850 and 1929 the number of planets varied depending on which publications you subscribed to. Most of the asteroids were removed, but the brightest three or four remained in many lists.

Then Pluto was discovered in 1930. It was quickly dubbed the 9th planet, putting the status of Ceres and the asteroids to rest once and for all. Remember, at the time, no one knew how different Ceres was from the other asteroids. If they could have seen it through better telescopes as a "disk" like Mars and Uranus, it would surely have remained a planet.

Interestingly, Pluto could not be resolved into a disk either. But somehow that didn't seem to matter anymore. The whole world was excited about a new planet being discovered. It was big news. There was an international contest to name the planet. A little girl in England won that competition.

It wasn't until 1994 that the Hubble Space Telescope was the first instrument to get a fuzzy disk-shaped image of Pluto. A few years later, adaptive optics would allow ground-based telescopes to get clearer images than Hubble.

In 1992, two years before Hubble took its photo of Pluto, the very first object beyond Pluto was discovered. Soon more were found, and a whole new class of "Kuiper Belt" objects was created.

When Eris was discovered in 2005, NASA immediately recognized it as the tenth planet. For more than a year, the Solar System was officially a ten planet system. Even it's discoverer, Mike Brown (who would later claim to be Pluto's Killer) agreed. He originally dubbed the tenth planet Xena, because Xena starts with an X, and X is the roman numeral for ten.

— Mike Brown, from his book, *How I Killed Pluto*

But that was only an unofficial temporary name. Persephone was offered as a permanent name for this planet, and that name was gaining popularity fast. Then the IAU stepped in and said (in effect), "Hold on, that's a Kuiper Belt object. We need to look into this," and started the whole debate about planethood once again

Yeah, about that "Kuiper Belt"...

Chapter Seventeen: Kuiper's Comet Zone

We should take a moment to be impressed by the mathematical wizardry of the astronomical community. They may have trouble with definitions, but they knock the ball out of the park when it comes to the mathematical part of their jobs.

Gravity is a funny thing; it never quite tapers off to zero. Every object in the solar system is the center of its own gravitational field, and every gravitational field affects the motion of every other body in the solar system to some degree. It is a complex four-dimensional field of moving force vectors. Astronomers and astrophysicists study these relationships, and they do it all from the perspective of a moving, spinning Earth. From here, they can tell where the bumps and lumps are in these fields— That's how they determine where the missing parts should be. That's how the outer planets were discovered. It's also part of the story for how the Kuiper Belt was discovered, including a multiplex of different zones within it. What the astronomers do is important. We would know nothing of these things without their amazing work.

But, as they say, no good deed goes unpunished. On with our task...

When it comes to naming insignificant things like the 79,058th asteroid, or defining terms that don't need to be defined, the AIU is very persnickety. But when it comes to naming the largest features in the solar system, like the Kuiper Belt, it seems that *no one is in charge*. Whatever the first name someone slaps on it happens to be, that seems to stick.

Who discovered the Kuiper Belt? If Pluto is a Kuiper Belt Object, as many astronomers insist, then Clyde Tombaugh discovered the very first KBO in 1930. It seems that he should get the credit. But at the time, no one understood there were more things out there to be found.

Pluto was a planet. End of story. Except, there was something weird about Pluto that eventually led to the "discovery" of the region named for

Gerard Kuiper. This weirdness bothered a lot of astronomers. They didn't like it.

Pluto was an outcast because its orbit is eccentric and tilted. For twenty years out if its 248 year journey around the Sun, it is closer to the Sun than Neptune. A few astronomers felt it just wasn't conventional enough for planet status. But the public was already convinced, and Pluto was useful as one more reason why we need to fund bigger and better telescopes. The first good picture we got of Pluto was via the Hubble space telescope.

Then, in the late 1980s, new technology made it possible to improve on that view using ground-based telescopes. Two astronomers, David Jewitt and Jane Luu, decided to use that technology to search the space beyond Pluto. There had been theories floating around about another planet out there. Jewitt and Luu didn't find that planet, but they eventually discovered the first Kuiper Belt object that isn't Pluto in 1992.

It should be noted that Pluto is by far the brightest object in the Kuiper Belt. It had been photographed many times before its 'discovery' in 1930; the earliest photo dates from 1909. But Tombaugh was the first to see it move and understand what that meant, so he got the credit, just like Herschel did with Uranus.

There are no records of any other Kuiper Belt objects being photographed before the 1992 discovery. After 1992, more and more Kuiper Belt objects were found and their orbits calculated. Many of those orbits were similar in size and shape to Pluto's orbit. The echoes of the Astroid Belt were ringing loud and clear to some astronomers. Pluto's orbit was looking less like the orbits of other planets and more like the orbits of other Kuiper Belt objects.

Astronomers also discovered that the Kuiper Belt is *huge*. It's not exactly a belt at all. It's more like a giant, fat donut. The thickness of the Kuiper Belt is about the same as the distance from Mercury's orbit to Neptune's orbit. Its volume is greater than the entire solar system inside the orbit of Neptune. And it's not composed of just one kind of orbit. The Kuiper Belt contains multiple classes of orbits sharing overlapping spaces.

Here's an abbreviated list:

Classic Kuiper Belt Objects (KBOs) are in a harmonious 3:2 synchronization with Neptune's orbit. This includes Pluto.

Scattered Disk objects (SDOs) are **not** synchronized with Neptune. This includes Eris.

Resonant Kuiper Belt objects resonate with Neptune's orbit in ratios different from 3:2.

Extended Region Objects intersect the Kuiper Belt but their orbits take them to the edge of interstellar space and back. This is where you'll find Sedna.

Neptune Trojans and *Centaurs* are asteroids and comets that bob in-and-out of the belt but also can be found in the lower regions around the gas giants.

Just as chaotic as the belt itself are all the names for the groups and sub-groups of objects within these systems. Those names include (but are not limited to): Plutinos, Plutoids, Cubewanos, Sednoids, Centaurs, 'Scattered Disk Objects', 'Extended Scattered Disk Objects', 'Detached Disk Objects'. and other things generally called 'Trans-Neptunian Objects'.

To recap: It's not a "belt". It's more of a vast region that's bigger than the whole classical Solar System. And it's not populated by a single class of objects like the Asteroid Belt. There are at least five different classes of orbits here, with more weirdness being discovered every year.

Now for more bad news. It is well known in the astronomical community that "Kuiper" is entirely the wrong name for it.

The "Kuiper Belt" was named for astronomer Gerard Kuiper, in spite of the fact that he wasn't the first to propose the idea, he didn't discover any Kuiper belt objects, and he believed it didn't exist. In fact, he was possibly the most vocal astronomer in asserting that it did not exist.

What we call the Kuiper belt should have been called the Comet Belt, or, if we insist on picking one of the many people who "discovered" it (other than Clyde Tombaugh), it should probably be called the Kenneth Edgeworth belt or the Fred Whipple belt or possibly the Julio Fernandez Belt. So why wasn't it?

Depending on how you define the word *discovered,* the honor could go to any number of people. There's a big difference between hypothesizing the possibility for something to exist, vs. doing the math to describe its potential size and position, vs. actually seeing it and taking photos.

In 1951, Gerard Kuiper speculated that "one or more small planets, like Ceres" might have existed beyond the orbit of Neptune in the early solar system. *— 1951, Astrophysics: A topical Symposium, McGraw-Hill Book Co., p. 402.*

(Notice the use of the word *planet* to describe Ceres. And this was in the 1950s—not the 1850s! Astronomers and their categories...)

Kuiper went on to say it would be puzzling if any objects persisted there to modern times. He didn't include a reference, but he may have been responding to earlier claims by astronomer K. E. Edgeworth *(1949 MNRAS 109, 609)* who was the first to propose that the outer regions beyond Neptune should be occupied by a large number of small objects, and could be a vast reservoir of comets.

Edgeworth had good reason to believe this. There's a problem with the comets. Comets are the mayflies of the Solar System. Relative to other things, they have very short lifespans. Their kamikaze dives into the inner solar system causes them to heat up and throw off gasses—that's where the tail comes from. Eventually they whither away to become dead, rocky asteroids. This can take a few million years, but given the age of the solar system, they should have all disappeared long before humanity came upon the scene to witness any of them. Something was replenishing them. There had to be a vast reservoir of new comets somewhere.

The working theory was that gravitational nudging from the planets would jostle this reservoir and cause new comets to gradually rain down to the inner solar system. That reservoir could only be at the outer edge of the known solar system.

Later, astronomer Fred Whipple, working with Astronomers S. E. Hamid and Brian Marsden, studied the gravitational effects a zone of comets would have on the rest of the Solar System. Their results were inconclusive.

Then in 1980, astronomer Julio Fernandez wrote a detailed paper about it. The paper is full of good science, and Fernandez arguably did more to bring the existence of this comet zone to common knowledge than anyone. Unfortunately, he didn't do his historical homework very well. The second sentence of the introduction in his paper gives undue credit to Kuiper for describing the belt.

Fernandez never actually uses the phrase "Kuiper Belt", instead he called it "a comet belt." The phrase "Kuiper Belt" was coined by Canadian astronomer Scott Tremaine. He was inspired by the first paragraph of the introduction of Fernandez's paper:

A comet belt located beyond Neptune has been suggested as a possible source for the observed comets. In an early version, Kuiper (1951) pointed out that such a belt would be the remanent of the outermost parts of the solar nebula, between about 35-50 AU, left behind after the formation of the planets.

Fernandez later admitted that referencing Kuiper was a mistake, and has suggested the name be retired in favor of something more appropriate. But momentum is hard to fight. Other astronomers followed Tremaine's precedent without doing their own research, and the name stuck, even though it has never been declared the official name by the IAU—the official naming authority of the astronomical community.

It seems that many authors typically do not do their historical homework, preferring to rely on the homework of those authors that they cite; such practice leads to a propagation of errors, sometimes where even the citation is in typographical error.

— www.icq.eps.harvard.edu/kb.html

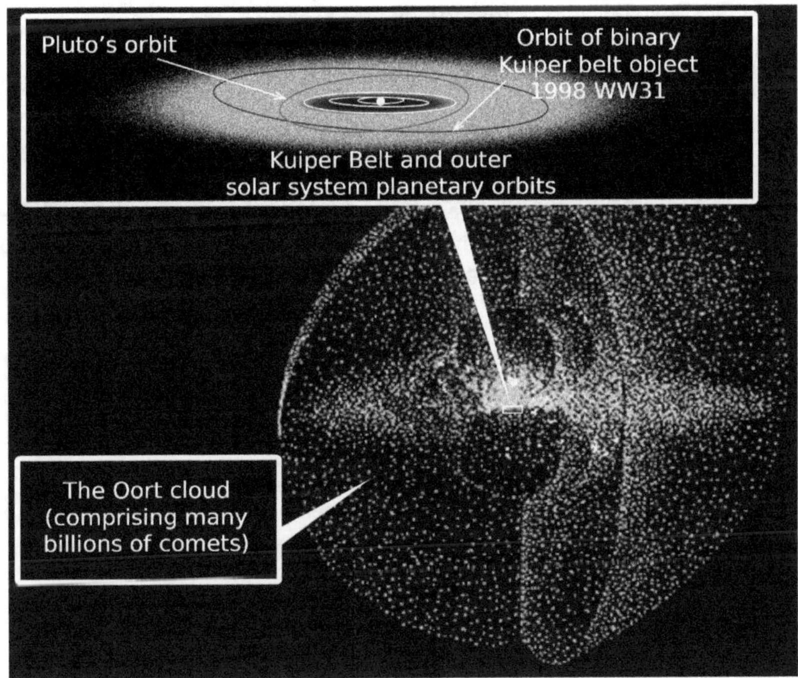

Since the IAU is in charge of naming things, and it is well known that the Kuiper Belt is mis-named, and the IAU isn't afraid of a little controversy or upsetting the public over reclassifications… then why not fix this unfortunate problem and put credit where it's due?

But then, 'where it's due' is another point of contention. Brian Marsden claims that Fred Whipple deserves the real credit. He said, "Neither Edgeworth nor Kuiper wrote about anything remotely like what we are now seeing. But Fred Whipple did." — from the book, *Beyond Pluto*, by John Davies, p 199.

Not all astronomers agree with Mr. Marsden though. The most common alternative to the name "Kuiper Belt" used in scientific papers is the "Edgeworth-Kuiper" belt. The only reason the word Kuiper is included is so the paper will be found by search algorithms when other scientists look for previous works to reference.

Kuiper was a prolific, hard working astronomer who was well liked, and he already has plenty of other things named for him. These include asteroid 1776-Kuiper, a crater on each of the planets Mercury, Luna and Mars, and an escarpment in Antarctica.

With that tidbit of history out of the way, we should all all understand that the two words *Kuiper Belt* are a placeholder for the "Tombaugh-Edgeworth-Kuiper-Whipple-Fernandez Trans-Neptunian Zone." I include Tombaugh in that list because he officially discovered the first object in this zone almost two decades before Edgeworth ever mentioned it.

(Let's digress for a moment. That's an awkward name. Since this region is so incredibly vast, and since there's no real agreement on who "discovered" it, or how to classify the objects within it, let's just call it what it is—the Trans-Neptunian Region of the Solar System.)

Ironically, the whole "Kuiper Belt" discovery is just a derivative of previous work anyway. A much larger zone that is much farther out was identified first, and it was also found by thinking about comets.

Astronomers classify the comets into two categories: short period and long period comets. Short period comets make their round trips in dozens to hundreds of years. Long period comets have round trips measured in thousands to millions of years. They have to come from somewhere far beyond the known Solar System.

In 1947, two years *before* Edgeworth mentioned a comet zone beyond Neptune, Dutch astronomer Jan Oort laid the groundwork for the existence of a reservoir of long period comets in a much farther region of the Solar System, a region we now know as the Oort Cloud.

But short period comets make their round trips too fast to come from way out there. They had to come from somewhere closer. The obvious answer is a zone just beyond Neptune. But there was no way to see anything that far away. Edgeworth, Kuiper, Whipple, Fernandez and others all kicked around that idea on paper. But the real goal of astronomers is to see it.

Some of the short period comets had to come from even closer orbits, perhaps somewhere between Jupiter and Neptune. Sure enough, when someone looked hard enough, that's what they found. Today these objects are known as the Centaur asteroids. The first confirmed Centaur asteroid was discovered in 1977 by Charles Kowal.

Kowal's asteroid was hanging out in an orbit between Saturn and Uranus. At the time, some astronomers considered it to be a new planet. Others wanted to know what an asteroid was doing so far away from the belt. The orbit was calculated and extrapolated, and it was determined that this new object's orbit was unstable. It crossed the path of Saturn, but without being in a harmonic orbit. Eventually it would be perturbed by Saturn and either dive into the inner solar system or be thrust away. This raised the question of how it had survived until today in a four-billion year old Solar System.

In time, more Centaurs were discovered. As of this writing (Sep. 2021) there are more than 750 known Centaur objects. I say 'objects', because some are called asteroids, some are called comets, and some are called scattered-disk objects, depending on who you ask. In 1989, Kowal's asteroid was seen sporting a faint tail. Its status was changed from asteroid to comet.

Centaurs are generally anything that's not a moon or a planet, zooming around somewhere between Jupiter and Neptune, and that crosses the orbit of at least one gas giant. But this is not a universally accepted definition.

Many of the centaurs have orbits that take them deep into the trans-neptunian regions. Many of these were the first trans-neptunian objects discovered, they just happened to be inside Neptune's orbit when they were first seen. Halley's comet is one of them.

The point is, the "Kuiper Belt" doesn't have a definite edge. It bleeds deeply into the solar system all the way down to Jupiter, maybe even deeper.

The very first object seen beyond the orbit of Pluto (that is not a star or galaxy) was discovered by Jane Luu and Dave Jewitt on Aug. 30, 1992. Brian Marsden, head of the IAU's Minor Planet Center gave it the designation 1992-QB_1.

About that designation— The Minor Planet Center uses a system that was devised in 1925. When a new object is reported, its sobriquet is based on the year, the half-month, and the order it was found within that half-month.

The first letter after the year is the half-month of discovery. Since there are 12 months, but 26 letters, I and Z are not used for that part of the name. The first half of January is A, the second half is B, the first half of February is C, etc. to the second half of December, which is Y.

The second letter after the year is the order of discovery, in alphabetical order. The first object found in the first half of January in 1980 would be 1980-AA. The second one found in that same half-month would be 1980-AB. The first thing found in the second half of January would be 1080-BA. When more than 25 things are found in one of these half-months, 1980-AZ is followed by 1980-AA_1, then 1980-AB_1, etc. That's how we got to 1992-QB_1. It was the 27th new object reported in the second half of August of 1992.

In case you thought something was funny about my numbers, the math is not wrong. The IAU omits letters I and Z for the first character of the designation. For the second character they omit I, but include Z. So their alphabet for the first character is 24 letters, and for the second character it is 25 letters. Why do they do this? No explanation was given.

With or without the letter I, the alphabet is just not big enough for modern astronomy. These numbers can get pretty large. There's an object with the designation of 2014-OS_{393}, and another one is 2007-TC_{434}, and another one is 2015-BZ_{517}. It goes on from there.

Something tells me astronomers enjoy making things more complicated than they need to be. Anyway, kudos to the prolific astronomers who rack-up these astronomical numbers!

Six months after 1992-QB_1 was found, Jewitt and Luu discovered a second object, called 1993-FW. By 1994 there were four more objects discovered. This new class of objects needed a name.

Apparently, it's a tradition in astronomy that a class of objects be named after the first member of that class. Following this logic, Brian Marsden called the new class of objects Cubewanos (think Q-B-1-ohs). But the name was not popular. Dave Jewitt started calling them Plutinos (little Plutos), since they have orbits that look similar to Pluto, and technically Pluto was the true first member of this class. (I'd like to know when this naming tradition started, since the Asteroids would have been more appropriately named Ceresoids if it existed in 1800. Unfortunately, I wasn't able to find an originating source for this 'tradition'.)

What lies between the trans-neptunian zone and the Oort cloud? No one knows. There is evidence for another large gravitational mass out there, but it hasn't been found yet.

And what about beyond the Oort? You didn't think that space was empty, did you?

Chapter Eighteen: The Free Planets

Way beyond Pluto, beyond the Comet Zone and the Oort Cloud, is interstellar space. And it turns out, there are planets out there.

Seth Shostak is an astronomer at the SETI institute in California. He and Iain McDonald, an astronomer from the university of Manchester in England, have been actively searching for rogue free floating planets (FFPs) in the space between the stars. Their conclusion: FFPs may outnumber planets in orbit around stars—by a *lot*.

"It's quite possible that the majority of the planets in the universe are untethered to stars." Shostak says. In other words, it has no orbit, unless you consider the galactic orbit around the black hole in Sagittarius-A to be its "star". How does this happen? More importantly, why is it important to the definition of a planet?

There are very few giant stars in the galaxy. Slightly more medium stars exist, like our Sun. But the vast majority of stars are red dwarf stars. The reason for this is not so different from the reason there are so few billionaires, and somewhat more millionaires, but the vast majority of people are living paycheck-to-paycheck every month. Becoming a giant star, like becoming a billionaire, requires a combination of starting in the right kind of environment and making all the right moves at the right times.

If the curve continues (and it almost certainly does), then there will be even more brown dwarfs and exponentially more FFPs than red dwarfs. But where do all these things come from?

It starts with a vast cloud of hydrogen, helium, and a smattering of other elements. Every particle in these clouds has its own gravity. En masse they combine to give the cloud a massive but widely dispersed, decentralized gravitational field.

Modern theory supposes that a passing star or a traveling shock wave disturbs the cloud, then parts of it start to move, then the cloud begins to collapse. Nothing in the universe is truly stationary, so the combined angular momentum of the particles causes the cloud to spin. Bands begin to shear from each other and coalesce. Chunks form in the chaos. Larger chunks are caught in each other's gravity. They either merge into an even larger chunk or smash each other into smaller pieces, depending on how much energy is in the collision.

This process continues until a swirling disk forms. At the center of this disk is the lump that will eventually become a star. The rest of the material will become planets or other stars and proto-stars within the system. A big enough cloud of gas can form whole families of stars.

But what if there's not enough matter in this disk to make a star? What happens to all the chunks of matter that get thrown out to interstellar space from the chaos of star formation?

If there's not enough mass to ignite the hydrogen into a fusion reaction, then you don't get a star. To do that requires about 80 times the mass of Jupiter. Instead of a star, you get a brown dwarf.

Brown dwarfs are not stars, but they are not planets either. They still generate heat and radiation (like Jupiter does). They can even have planets of their own (like Jupiter does). The difference is that they can't fuse hydrogen (the definition of a star). Instead, they fuse deuterium.

Deuterium is just hydrogen with an extra neutron in its nucleus. This extra neutron makes it twice as heavy and a little less stable than ordinary hydrogen. There are roughly 26 deuterium atoms for every million hydrogen atoms in the universe. It's pretty rare.

Below about 13 times the mass of Jupiter even deuterium fusion stops happening. Thirteen times the mass of Jupiter is the boundary between brown dwarfs and gas giants, regardless of anything to do with orbits at all.

Somehow gas giants can still generate a lot of heat and radiation, enough to last billions and billions of years. Jupiter's moons Io, Europa and Ganymede all receive more radiation from Jupiter than they do from the Sun. (Callisto is far enough away from Jupiter that it receives less total radiation, from Jupiter and the Sun combined than Earth does from the Sun alone. We should go to Callisto sometime. The view of Jupiter and all the other moons in the nighttime sky would be awesome.)

If you move the whole Jovian system to interstellar space, it would be its own little solar system, but with a free floating 'planet' instead of a star. I have to put the word planet in quotes, because in that situation it's not technically a planet. It's not technically a star either. The term *planetar* has been suggested, but it never caught on. Black Dwarf was also suggested, but this term is reserved for a white dwarf that has lost all of its heat, a situation that has been calculated to be impossible within the current age of the universe. Substar is another term that has been suggested. Technically, the most accurate term for one of these objects is *asteroid.* They are genuinely star-like, but not stars. They can even support life in the same way stars do. It is theoretically possible for life to exist on Europa from the energy it gets from Jupiter. Maybe the IAU can redefine all the asteroids to be Ceresoids (similar to the Plutoids and Sednoids), then we can free-up the term Asteroid for the free-floating rogue planets that seem to outnumber the stars.

As we move farther down the scale, there could be many, many, many more Earth-sized and larger rocky planets floating between the stars than orbiting them. These are free floating *planets*. Even thought they do not fit in the IAU's flawed definition.

Chapter Nineteen: Asteroids, Meteors, and Comets

In the ancient world there were only four kinds of things in the sky—the stars in their fixed pattern, the things that moved and were persistent (planets), the things that moved and only lasted a few weeks or months (comets), and things that moved and lasted only seconds (meteors).

The meteors were known to be an atmospheric phenomenon. They share the same root word with meteorology—the study of atmospheric phenomena.

The comets were more mysterious. Instead of lasting only seconds like the meteors, they remained in the skies for weeks or months. Today we know that meteors and comets and asteroids are all essentially the same thing—chunks of rock and ice, in motion, in space. The difference is their temporary situations.

Most meteors burn up in the atmosphere. Some hit the ground and become part of the Earth. Comets also "burn-off" their volatiles over time, becoming less ice and more rock as they do. In both cases, they are in an accelerated rate of transition.

The words were originally coined to describe what is *happening*. They don't work as well for describing what the objects *are*. Comets are generally more ice that rock. So is Rhea, the second largest satellite of Saturn. So is Pluto, if measured by volume. So are Uranus and Neptune, commonly referred to as the "ice giants". But none of those things are comets.

Is a comet without a tail still a comet? Many of the mostly-ice objects without tails go by the name of centaur asteroids and TNOs. Right now, Halley's comet is sailing past Neptune on its way into the Kuiper Belt. In the year 2023 Halley's comet will be at its farthest point from the Sun, solidly within the trans-neptunian bounds of the "Kuiper Belt" and sporting no tail.

The interstellar traveller known as Oumuamua passed close enough to the Sun to be a comet, but it never developed a tail. As it flew away from the Sun it did a surprising thing—it accelerated. There are two popular theories to how it did this. The first one is that radiation from the Sun caused the object to outgas on one side, providing thrust. If this is the case, then it was a comet with a very powerful but also invisible tail. Since we've never seen this behavior from any other object in the Solar System, the second most popular explanation is some kind of alien technology. Maybe a solar sail of some kind. Unfortunately, Oumuamua is gone now, so we may never know.

To grow that tail requires having a lot of volatile ices and a very low mass so the gasses can escape when the ice sublimates. This requires being close enough to the Sun to get warm enough for sublimation. The less volatile the composition of the comet is, the closer it has to be to the Sun for a tail to form. That's really the only difference.

Comets and meteors are not really 'things,' they are more like events in the life of an asteroid.

Most asteroids may not even be in the asteroid belt. It's just a place where the density of asteroids is extremely high. Some astronomers believe there are more asteroids in the wake of Jupiter's orbit than in the asteroid belt. If you group together the two sets of trojan asteroids in Jupiter's L4 and L5 positions, there could be more asteroids there than in the main belt, and this doesn't count the unknown quantity of asteroids in the outer regions of the Solar System.

Regardless of where they are, all the asteroids and comets in the Solar System share one important quality. They are all smaller than Ceres.

Getting to the point of this chapter—the single biggest reason Pluto was demoted from planethood was because of a pattern established with the asteroid belt and Ceres. Modern astronomers were following a precedent set 150 years earlier.

Whether a rule is good or bad, people love to follow precedent. It's easy. You don't have to think too hard about it. And it eschews responsibility when you can point to the precedent and say, "We were just following precedent…"

If you don't follow the precedent you take on a lot of responsibility. You become vulnerable to criticism. The old idea must be replaced with something new. Creating something new and original is hard. Creating something original that's also good and useful is doubly hard.

When a population of objects was discovered in the space beyond Pluto, the asteroid belt naturally came to mind. The "Kuiper Belt" was sometimes described as a second asteroid belt in our solar system. It would later be found to be profoundly different, but the image of an "asteroid-like region" was already established.

It was the pattern of Ceres's reclassification from planet to asteroid that led to Pluto's reclassification. It was wrong then, and it is wrong now.

When the asteroids were discovered, no one knew how big they were. No one knew what shape they were. No one knew what they were made of. Something weird was going on here, but what it was, no one could tell.

They were given the most inappropriate name. Asteroid, aster-oid, is latin for 'star-like'. And yet, there is absolutely nothing star-like about the asteroids. They are small. They are rocky and/or icy. They are cold. They emit no radiation. They have no magnetic fields. They do not shine. So how did they get that horribly inappropriate name?

It was a name borne of ignorance. Astro-photography did not exist yet. All star-charts were drawn by hand, with pencil and paper. The technology of the time gave them no information other than the position, the relative brightness, and the path of these points of light in the sky.

It was decided that there were too many planets inhabiting the same orbit. Compared to the previously established planets, these new things were as dim as many of the stars. No matter how far they zoomed in, these points of light never resolved into a disk like the other planets. So the term asteroid, meaning "star-like", was given to them.

Keep in mind the times. No one knew what the stars were made of. The first use of spectroscopy to analyze the elements in the stars and finally answer the question of what stars are made of happened in 1840 by William Higgens, thirty-eight years after the name asteroid had been applied to the band of rocks floating between Mars and Jupiter.

Remember that 'asteroid' was originally intended as a temporary label, only to be used until the true nature of these strange new objects could be determined.

Neither the appellation of planets nor that of comets, can with any propriety of language be given to these two stars ... They resemble small stars so much as hardly to be distinguished from them... From this, their asteroidal appearance, if I take my name, and call them Asteroids; reserving for myself, however, the liberty of changing that name, if another, more expressive to their nature, should occur.
— William Herschel, 1802. On the observations of Ceres and Pallas

I think we're a little overextended on changing the name to something more expressive of their nature.

It's clear Ceres was a planet all along.
— Alan Stern, in a 2010 interview with NBC News.

By the mid 19th century it was clear these were not "star-like" things, but tiny planetoids or planetesimals. The German publications seem to have been first to try and re-classify them, calling them 'Kleine Planeten' (tiny planets). Soon the French were calling them 'Petites Planetes'. But leave it to the Americans to screw things up. The U.S. Naval Observatory continued to call them 'asteroids' until 1868, when they finally joined the party and renamed them 'small planets.' Then in 1892 they reverted back to 'asteroid'. They switched to calling them 'minor planets' in 1900, then back to 'asteroid' in 1929.

Even the IAU tried to patch the issue. Until recently the IAU referred to the asteroids as "minor planets". But in the IAU's Resolution B5—the same one that issued the definition of a planet and said that dwarf planets are not planets—also states: "All other objects, except satellites, orbiting the Sun shall be referred to collectively as 'Small Solar System Bodies'."

I suppose that's SSSBs for short. Yeah, that'll catch on.

It looks like we'll be stuck calling them asteroids for a while.

The original discoverers of the asteroids had no way to see the glaring differences between Ceres and the rest of the bunch. If their telescopes had been able to see that Ceres was uniquely spherical, and that Ceres contained one-third of the mass of the entire belt *combined*, therefore utterly dominating it... I suspect they would have retained the classification of *planet* for Ceres.

But why do the asteroids exist in a concentrated belt formation in the first place? If Jupiter did not exist where it does, or if it were the size and mass of the Earth, for instance, then the asteroid belt would have coalesced into a planet and swept its orbit clean long, long ago. Instead of an asteroid belt, we would have one singular planet instead.

Ceres is not a full-fledged planet because Jupiter's massive gravity keeps stirring things up and disrupting the process of formation. You can actually see the knife-cuts Jupiter makes in the belt. They're called Kirkwood gaps. They are regions where the asteroids get divvied up and split into separate groups due to the orbital harmonics of Jupiter's immense gravity. The gray areas in the graph below represent the number of asteroids at that orbit. Ceres is located at approx. 2.75 AU.

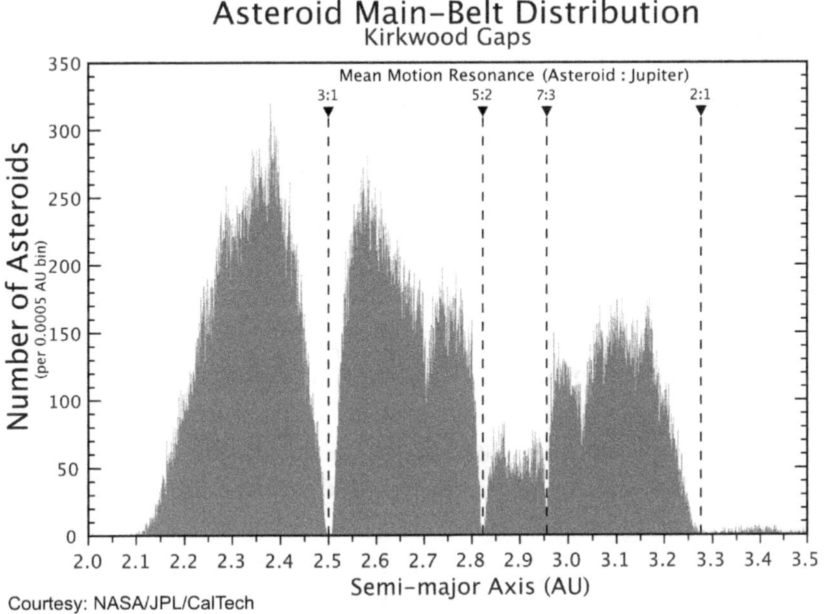

Courtesy: NASA/JPL/CalTech

Imagine if you were not allowed to join the country club because there was trash on your lawn. You've got all the other requirements met, but some nasty neighbor keeps trashing your yard, every single morning.

Yeah, that's Ceres' situation. It's a victim of Jupiter.

Chapter Twenty: Is the Moon a Planet?

Is it a moon or a planet? Most people would say our moon is not a planet. But if the Earth disappeared, could the moon qualify as a planet on its own? It is definitely round, but would it be a planet or a dwarf planet?

Of course, this is a pointless exercise since the Earth is not going to disappear and the Moon is a satellite of Earth. That automatically disqualifies it.

But then again, that depends on how we define the word 'orbit.' Does the Moon really orbit the Earth? The short answer is yes, but that doesn't tell the whole story. It's like asking if the Harry Potter series of books is about the relationship between Ron Weasley and Hermione Granger. Yes, that's a big part of it, but it's not the *main* story. Likewise, there is a much bigger story behind the Sun/Earth/Moon relationship.

First, let's get this out of the way— The word 'moon' is problematic. Is it a category or a thing? There are 'moons' of Mars, for instance. And moons around every one of the classical planets except for Mercury and Venus. According to the IAU, there is only one Moon in the Solar System. All the other objects previously called 'moons' should now be called 'natural satellites.' Where's all the fuss over people calling Phobos and Deimos moons? It's no different from the fiasco around the word planet.

If we give-in to the common usage, and continue to use the word 'moon' for the category of natural satellites, then we should stop referring to our moon as *the* Moon and start calling it by its proper name, Luna. We sent astronauts there in Lunar landers to explore it and collect Lunar samples, not a moonnar lander for moonar samples. We say that the ancient Romans and modern Muslims used a Lunar calendar, not a moonar calendar. Someone who is exhibiting the lunacy of a lunatic is said to be under the influence of Luna, not Moona.

Luna is the moon of Earth, like Phobos is a moon of Mars. Callisto is a moon of Jupiter. Charon is a moon of Pluto. We call ours *the* Moon because Earth is the only planet with exactly one moon (this is debatable, but we'll deal with that later).

Calling it Luna will also be more appropriate once we prove its status as a planet too. It would be too painful to have a planet with a name like "the Moon" in the Solar System. But that's getting ahead of the story.

No one knows exactly how many moons there are in our Solar System. Not all of the ones we know about have names. There are asteroids with moons and comets with moons too. Some moons are bigger than the smallest planets. Some the size of pebbles. There is no defined lower limit.

Our moon is not like any other moon in the Solar System. It is far more than just a moon. It should be considered a proper *planet*. That's not my opinion. It's a widely held belief among scientists, including notable names like Isaac Asimov.

We might look upon the Moon, then, as neither a true satellite of the Earth nor a captured one, but as a planet in its own right, moving about the Sun in careful step with the Earth.

To be sure, from within the Earth-Moon system, the simplest way of picturing the situation is to have the Moon revolve about the Earth; but if you were to draw a picture of the orbits of the Earth and Moon about the Sun exactly to scale, you would see that the Moon's orbit is everywhere concave toward the Sun. It is always "falling" toward the Sun.

All the other satellites, without exception, "fall" away from the Sun through part of their orbits, caught as they are by the superior pull of their primary, but not the Moon.

— Isaac Asimov, from his book, *On Time, Space & Other Things*

If you look hard enough you can find references to Earth and Luna as sister planets in a dual-planet system going back to the late 1800s. The International Journal of Astronomy and Astrophysics (Vol. 7, No. 4, Dec. 2017) contains a hard scientific paper by David G. Russel, proving that the Moon meets all the criteria of the IAU's definition for a planet.

Remember Jean-Luc Margot's rigorous, mathematical definition for a planet's capacity to "sweep the neighborhood"? His formula works beautifully to confirm the eight classical planets, and also to exclude Pluto and the other dwarf planets. But there's a problem. When you apply his formula to Luna, it qualifies as a full-fledged planet. Not a *dwarf* planet.

This proves that Luna satisfies the IAU's third criteria for planethood. It can clear its neighborhood. Luna is clearly round, so it satisfies the first criteria too. That only leaves the second criteria; in orbit around the Sun. Let's take a closer look into that one.

An orbit is a gravitational link between two objects, like the Sun and the Earth, or Earth and Luna. But let's not forget that the Sun also has an influence on Luna. The fact is, when you do the math, the Sun's gravity has more than twice as much influence on Luna's path than Earth's gravity does. *No other satellite in our solar system is like this.* For every other natural satellite, the host planet is its major gravitational influence. Everything else, including the Sun, is a distant second.

But not Luna. Our moon only appears to orbit the Earth because of our ground-based perspective. When viewed from a distance, the paths of Earth and Luna are more like a lane-changing race around the Sun.

Let's consider the Martian moons for a moment. Imagine you're standing somewhere near the equator of Mars. A day for you is just a little longer than 24.5 hours. At the moment of this experiment, the Sun has just dipped below the horizon. It's the beginning of the Martian nighttime.

You look up to find Mars's two moons, Phobos and Diemos. Phobos is the larger of the two moons. It is 13.8 miles across and whips around the planet at a speedy 7.66 hours per orbit, so it's easy to find. Deimos is a petit 7.8 miles across and a bit farther away. Deimos takes a leisurely 30.35 hours to make one orbit.

Do you see the problem? Even though both Phobos and Deimos travel the same direction around Mars, one is traveling faster than the planet spins on its axis, and the other moon is traveling slower than the spin. Deimos takes longer than a Martian day to go around, while Phobos can make more than three orbits in a single day. To an observer on the surface of the spinning planet the moons will appear to move in opposite directions. But this is only an illusion. You have to be far away from Mars and its moons to see the truth.

What does this have to do with our moon? Let's make that drawing Isaac Asimov mentioned and take a look.

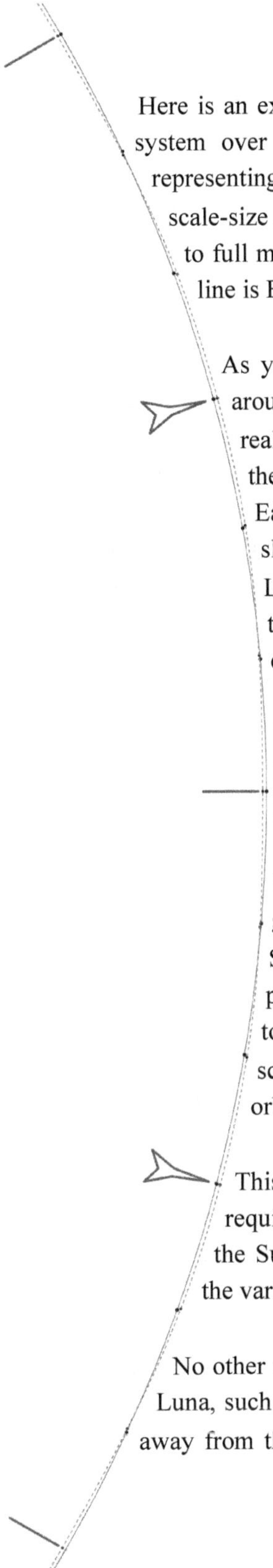

Here is an exact scale diagram of the actual orbits of the Earth-Moon system over two months (the orbital paths are to scale. The dots representing the Earth and Moon are ten times larger than actual scale-size for the sake of printability and visibility.) The arrows point to full moons, and the lines point to new moons. The solid curved line is Earth's path, and the dotted curved line is Luna's path.

As you can see, our Moon only appears to move in a circle around the Earth from our perspective on the ground. What's really happening is this—when Luna is closer to the Sun than the Earth is, it goes faster than the Earth. When it passes the Earth, it swings out to a larger orbit where it slows down, slower than the Earth's orbit. Then, as the Earth pulls ahead, Luna swings back to the inner path and does the whole thing over again. The two planets take turns overtaking each other on their race around the Sun.

This only works because the Earth's gravity is causing Luna to change lanes every time they pass each other. In that sense, it is true to say that Luna orbits the Earth.

But it is also true that the Sun has more than double the gravitational influence over Luna than the Earth does. The Sun/Luna relationship is the main story here. Luna's primary orbit is around the Sun. Its orbit is always concave towards the center. Luna's orbit around the Earth is scientifically, mathematically proven to be secondary to its orbit around the Sun.

This means our Moon satisfies ***all*** ***three*** of the IAU requirements for being a planet. It is round, it is in orbit around the Sun, and it has enough mass to clear its orbit according to the various equations formulated to calculate this.

No other natural satellite can do this—not even the ones larger than Luna, such as Titan, Ganymede, Callisto, and Io. They are all too far away from the Sun to be able to clear their orbits on their own. (But relocate them to anywhere between Mercury's orbit and Earth's, and they could do it just fine.)

Here's another way to visualize it— If Earth disappeared, Luna would continue on its path around the Sun pretty much just like it does today. But if Jupiter suddenly disappeared, all of its moons would scatter like marbles dropped on a hard floor. This goes for all the other moons in the Solar System too.

Have we missed anything? I'm sure someone will point out that Luna can't be a planet because it hasn't cleared the Earth out of its orbit. In that case, the Earth can't be a planet either, because it hasn't cleared Luna out of its orbit. Since Luna's primary gravitational bond is to the Sun, it cannot be discounted as just a satellite of Earth.

One method for testing your logic is to reduce it to an extreme case. Let's do that. Let's imagine there are only two months per year instead of thirteen (it's true, there are 13 of the 28-day lunar cycles in 365 days, with one day left over. We should fix our calendar too, but that's a different book...)

In the case of a two month year, the orbit of our moon would look something like this diagram at right:

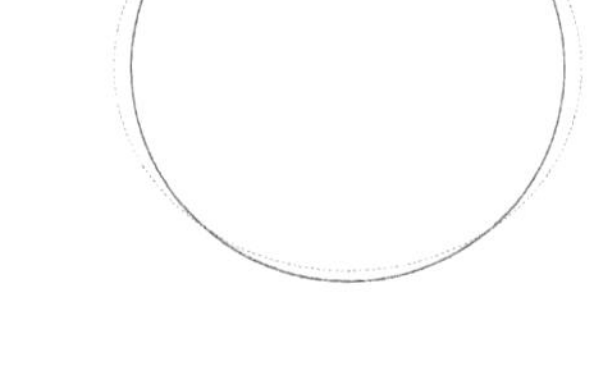

Again, the solid line is the orbit of Earth, and the dotted grey line is the orbit of Luna.

Here the moon's orbit looks more like a proper ellipse rather than a rounded polygon. It's easier to see how it orbits the Sun but also weaves in and out of the Earth's orbit.

Now let's reduce it again to see what one month per year looks like, and let's loosen-up the orbit a bit too.

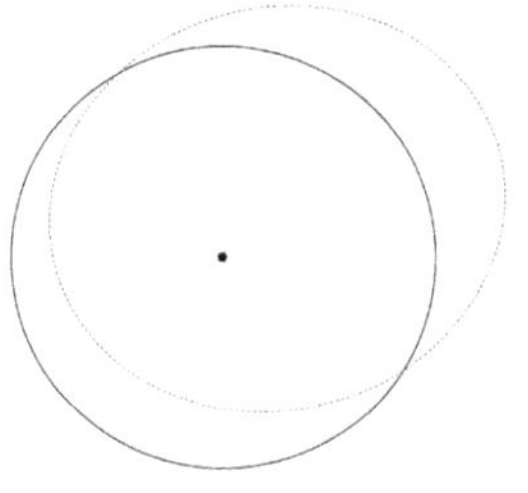

On the left is the orbit of Earth in the dark line, and the orbit of an asteroid named Cruithne in the grey dotted line. It's orbital duration is equal to Earth's. They share the same year.

Cruithne is gravitationally bound to the Earth in a 1:1 orbital resonance, but like Luna, that's secondary to its stronger gravitational link with the Sun.

But is it a moon of Earth? Some astronomers say it is. Most say it's not. Each has their own logical reasonings. One thing is for sure, Cruithne is not unique. It's one member of a small class of objects that are in orbit around the Sun but also gravitationally bound to Earth.

They are known as the Aten class of asteroids, but don't worry, they won't crash into us. They have a resonant orbit with Earth that keeps them away when they make their orbital lane changes. Just like Luna does. Either they are moons of Earth, or they are debris in Earth's not-so-clear orbit.

Here's an interesting side note to segue into the next chapter: When viewed from the perspective of the Earth, Cruithne's orbit traces the shape of a horseshoe in our nighttime sky. Astronomers call this a "horseshoe orbit," event though we know it makes an ellipse when viewed from a distance.

It seems that sometimes it's okay to classify things by simple appearances ("horseshoe orbits" that aren't actually horseshoe shaped, a moon that isn't actually a moon) but other times it's not okay (Pluto is round and orbits the Sun, but...)

To avoid the controversy over whether these are moons or not, most astronomers call them "quasi-satellites" and leave it to the rest of us to wonder what that really means. There is no definition for what is, or is not, a qualifying orbit for a moon around a planet.

Luna is not a moon, but Cruithne and a few dozen other Aten-class asteroids might be. It all depends on how we define an orbit.

Believe it or not, these are not even the most interesting examples…

Chapter Twenty-One: Let's Talk About Orbits

The IAU's definition for a planet depends heavily on the word "orbit", but they offer no definitions for what an orbit is, how wide it is, what's the right kind of orbit, or how to properly attribute an orbit's primary anchor.

If orbits are important, then we need to make sure we get it right.

As we've seen, orbits can be deceiving. Luna appears to orbit the Earth, but the math proves that it orbits the Sun first, and the Earth second. Without clarification on how to attribute an orbit, Luna must be considered a planet.

Luna is unique in this respect. But there are other unique orbits too. One of the most interesting orbits in the solar system is a single orbital path shared by two moons of Saturn. Say hello to Epimetheus and Janus.

Two Moons, One Orbit

These two moons share the same orbit, but they do not orbit each other, and they do not travel around Saturn at the same speed. Yet they never crash into each other either. They employ a kind of gear-shifting trick to get away with it.

I said "the same orbit", but that depends on how exacting you want to be. The diameters of their two orbits are different by only 50 kilometers, less distance than the width of either moon. Janus's smallest dimension is 152 km. Epimetheus's smallest dimension is 106 km. It is impossible for them to pass each other without colliding.

Their orbits are almost perfectly circular, and they travel around this circle at different speeds. So why don't they crash?

Whichever moon is in the tighter orbit around Saturn goes a tiny bit faster than the other moon. This is just how orbits work. The closer you are to the center of rotation, the faster you go.

When the faster moon starts to catch-up to the slower moon, their mutual gravity pulls them toward a common middle-ground. That 50 kilometer difference in their orbits goes to zero. Then inertia carries each moon a past the middle. They overshoot the center long before they have a chance to collide. Now the slower moon speeds up to pull away, and the faster one slows down.

These changes in speed are minimal. Their orbits differ by about a half-minute for every full orbit. It takes roughly four years between each near-miss event.

It's not a big enough shift to constitute a 'lane change'. Neither has moved enough to be able to pass the other one. Instead, this interplay of gravity and momentum has an effect similar to switching gears. One upshifts while the other downshifts.

It is theoretically possible for two planet-sized objects to share an orbit around a star in this way too. Their orbits are not clear, and they are not moons of each other. Are they planets? Yes, of course they are. They just don't fit the definition.

Lissajou Orbits and Lypunov Stability

In an earlier chapter we saw that things can orbit around islands of gravitational stability known as the L4 and L5 positions. Here's that diagram again showing where they occur. There are only two masses in this diagram. Everything else is empty space, including the points labeled L1 through L5.

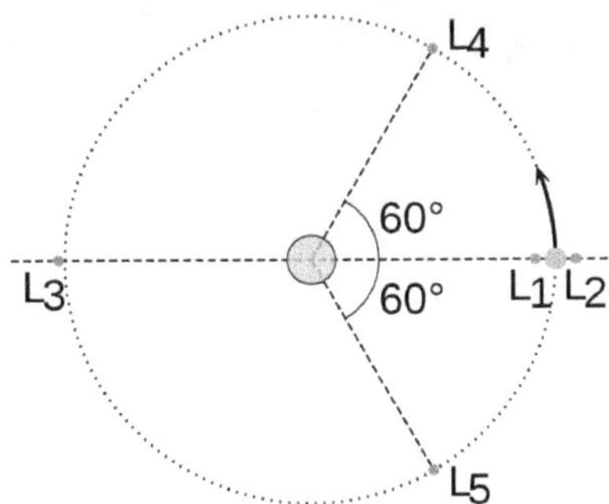

Orbits around the L4 and L5 positions are called Lissajou orbits. These orbits tend to be shaped like a tadpole, or a bean. They don't just look that way as seen from Earth, they really are that shape. They are not ovals or ellipses or circles. They are actually bean shaped. Objects in Lissajou orbits are generally called Trojan Asteroids. Earth has one Trojan asteroid that we know of. Object 2010-TK_7 orbits the Earth/Sun L4 position. Here is NASA's map of its orbit over a year:

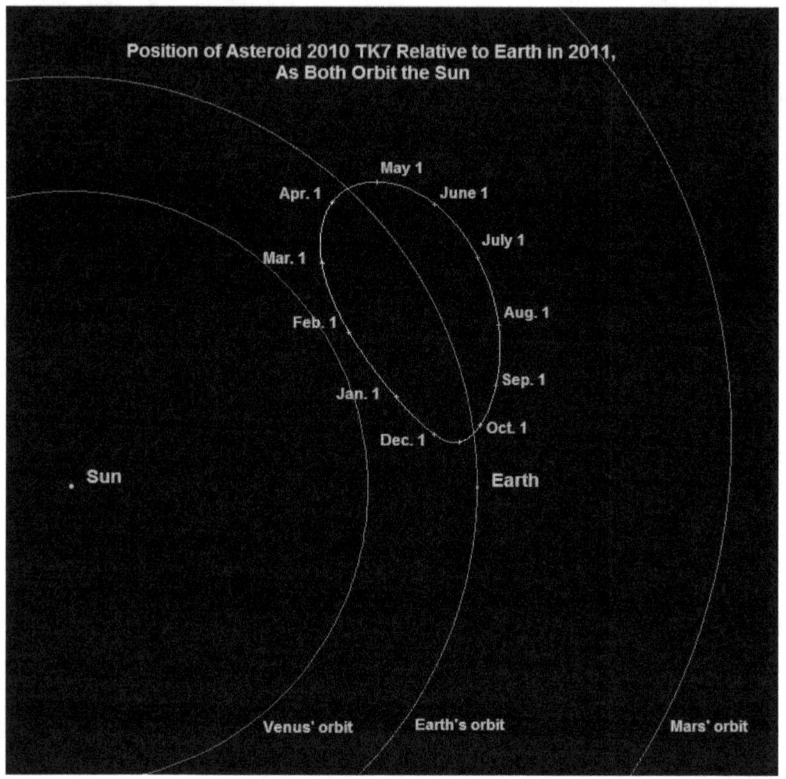

Earth Trojan 2010-TK_7 and its bean-shaped orbit around the Earth/Sun L4 position

A special kind of ultra-stable Lissajou orbit is called a Lypunov orbit. Lypunov stability occurs when the orbit is exactly in the plane of the two larger masses.

There can be multiple objects in orbit around the same L4 or L5 positions. Mars has four known Trojan asteroids. There are hundreds of thousands of known objects orbiting the Jupiter/Sun L4 and L5 positions. Some astronomers estimate there may be millions of unknown objects there—more than in the main asteroid belt.

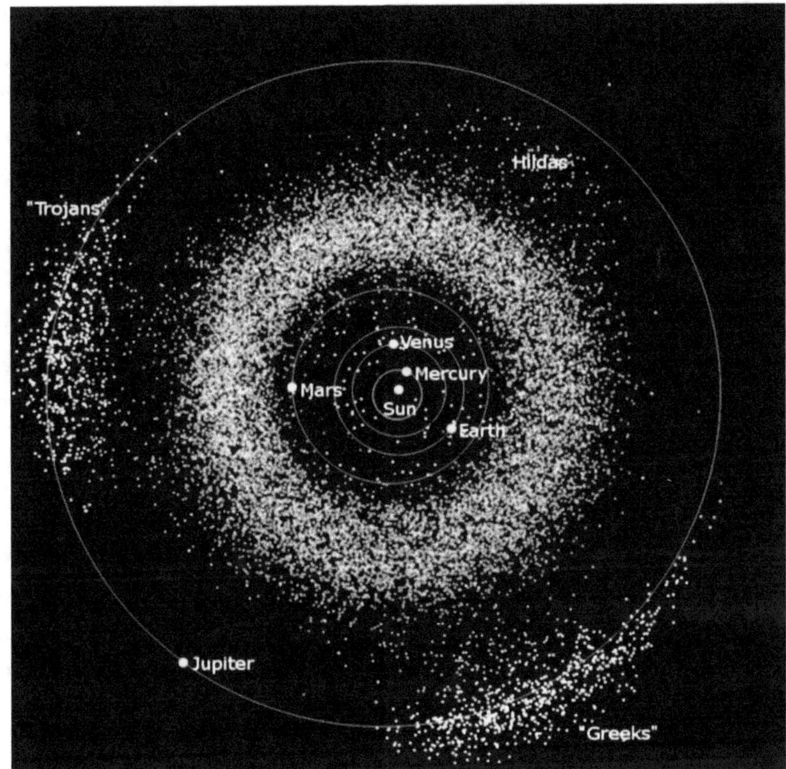

Jupiter Trojans and the asteroid belt

But once again, Saturn has the most interesting party going on.

Two of Saturn's moons (Tethys and Dione) have objects orbiting their L4 and L5 positions with Saturn. Which raises the question: since these objects are in a Lissajous orbit between Saturn and one of its moons, are they trojan asteroids, or moons of Saturn, or something else?

Let's scale that scenario up and play with it. Since binary stars are so common in the galaxy, do planet-like objects exist within the Lagrange points between giant stars and red dwarfs? And since their orbits will not be 'around their Sun', can we call them planets? Which star would they be planets to?

This is a very plausible scenario. As previously discussed, the Alpha Centauri system has two stellar masses near its center and a smaller red dwarf star circling farther out. There could be a planetary body orbiting the red dwarf's L4 or L5 points.

It gets better — you might have noticed that the 360 degrees of a circle is evenly divisible into six, 60-degree arcs. What would happen if you put another Earth at the Earth-Sun L4 and L5 positions, then two more at their L4 and L5 positions, then another to complete the hexagon?

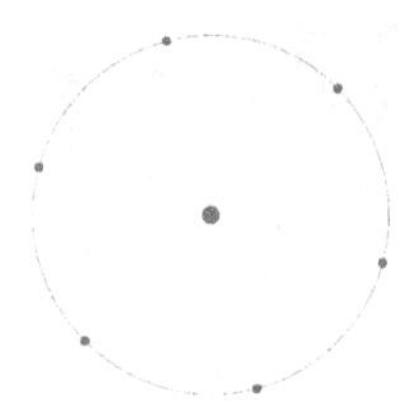

This is known as a Klemperer Rosette. For complicated gravimetric reasons, it's unstable. But strangely, seven planets in such a rosette do seem to be stable. This is virtually impossible to occur naturally, but a sufficiently advanced civilization might be able to construct one. If we ever discover something like that, maybe we can ask them how they define a planet.

(They'll probably say something like, "Whatever the locals say it is.")

By the way— The current theory for how our moon formed is related to this idea. A planet named Theia formed in one of Earth's L4 or L5 positions. It grew too big (about the size of Mars) and its orbit became unstable. Then it broke loose and "fell" into the Earth. The resulting cataclysm spewed enough ejecta into orbit around the new planetary amalgamation to eventually clump together into our moon.

> To see some outstanding simulations of Klemperer Rosettes and more, I highly recommend visiting this amazing page full of orbit simulators. It even includes the source code!
>
> **http://burtleburtle.net/bob/physics/kempler.html**

Halo Orbits

Orbits around the L1, L2 and L3 positions are called Halo orbits. These orbits are not stable over the long term, but they are useful parking spaces for a spacecraft that can make periodic adjustments. We have a solar observing "satellite" at the Sun/Earth L1 position. I put the word satellite in quotes because it doesn't actually orbit the Earth, so it's not really a satellite. But there's no better name for it than that.

Long ago it was theorized that a second Earth could exist in the L3 position on the opposite side of the Sun. This is known as the counter-Earth theory. In 1969 it was the subject of the movie *Journey to the Far Side of the Sun,* where an astronaut visits the other planet to find it's almost identical to ours, but with a critical difference… (It's a fun movie.)

Alas, since the L3 position is not long-term stable, there's no chance for another Earth to be hiding out there. But we've had a look anyway. NASA sent probes into polar orbits over and around the Sun for a variety of reasons, but found nothing hidden in our L3 position.

Irregular Orbits

Our moon, Luna, is a strange beast. It looks like a sphere, but if you could see gravity, it would look like a lumpy, bumpy glob. All planets have variations in their gravitational fields, but none is as bumpy as Luna's.

It's because of the Mascons. No, that's not the name of an alien species. It's short for mass concentration: mas-con. Beneath those smooth grey areas on Luna's surface are huge lumps of dense mass, probably iron, and probably the result of giant asteroid collisions long ago (or collisions with other mini-moons).

Whatever they were, they were embedded just below the surface, then engulfed in magma to create a deceptively smooth region of unusually high density.

Other than four very specific paths, there are no stable orbits around our moon. Any other orbit will degrade and crash into Luna's surface within a few years, or sooner.

Or will it?

James Meador is a researcher at the California institute of Technology. He has calculated that the abandoned Apollo 11 ascent stage could have oscillated back and forth between a series of unstable orbits with different eccentricities, different altitudes, and different orientations to the Lunar surface, creating an unlikely pinball-like irregular orbit around the planet that has lasted more than fifty years. Unfortunately, this has not been proven yet, but it sounds like an amazing project for a future astro-archaeologist!

Like Epimetheus and Janus, the math of gravitational systems demonstrates that some very odd orbits are possible. If we're going to define objects for the kinds of orbits they have, then we need to account for these kinds of situations too.

Other Moons That Aren't Moons.

If a moon is a natural satellite of a planet, what should we call the natural satellites of asteroids and comets? They do exist. As of September, 2021 there are 433 of them, with more being discovered all the time.

So far, no one has discovered any natural satellites orbiting any of the moons in our Solar System, but they are mathematically possible. Our Sun orbits a black hole at the center of the Galaxy (a first-order orbit), and Mars orbits the Sun (second-order), and Phobos orbits Mars (third-order orbit).

Having moons of moons (fourth-order orbits) may just be a matter of scale and/or luck.

By the way, this idea of *first-order* and *second-order* orbits is not part of any official nomenclature I can find. There doesn't seem to be a firm standard for classifying primary and secondary orbits. For instance, the planet orbiting Proxima Centauri is sometimes called Proxima Centauri B, and sometimes Alpha Centauri Cb. If that planet has a moon, what would its designation be? And what if that moon has its own moon…?

Retrograde Orbits

Most things in the Solar System go around the Sun in the same direction. In the case of moons, most of them go around their planets in the same direction too. This is called *prograde* motion. It seems to be a common property of all solar systems. When viewed from above the Sun's northern pole, most things in our Solar System orbit in a counter-clockwise direction.

I said "most things" because some things go the opposite direction. These orbits are called *retrograde* orbits. The most famous thing with a retrograde orbit is Halley's Comet. Many of the smaller moons of the gas giants are also in retrograde orbits. The largest object in retrograde orbit in our system is Neptune's moon Triton, the seventh largest moon in the Solar System.

Does it matter? Yes. A moon in retrograde orbit around a planet is usually more stable than a prograde orbit. Ironically, most of the moons in our Solar System are in prograde orbits. Does this mean we're in danger of losing a few moons? Yes. It does. In fact, moons come and go all the time. Mars may have had many moons in the past. Its inner moon is gradually getting closer to the planet and is destined to crash into it in the future. Mars's outer moon is slowly moving away. It will eventually escape.

This is true of our own moon too (if it is a "moon"). A few billion years ago it was only one quarter as far away from Earth as it is now. Every year it gets a little farther away. Eventually we will lose it.

Don't worry, this won't actually happen. Our Sun will expand and engulf both Earth and Luna long before our moon will get away. But either scenario would be bad for Earth. If Luna broke away, it would still orbit the Sun. We'd either have a co-orbital situation like Janus and Epimetheus, or Luna would eventually crash back into Earth, like planet Theia did four billion years ago.

Retrograde moons are usually more stable, but in systems with multiple prograde moons, the retrograde moons can become less stable (orbits are complicated…) In the case of the retrograde moon Triton, it contains 99.5% of all the mass of all the moons of Neptune combined. Triton is the Jupiter of the Neptunian moon system. Its orbit is safe.

The point of all this is that orbits can, and do, change. Orbits change more often and more dramatically than planets do. Computer models of the evolution of our Solar System suggest that Jupiter has migrated to several different orbits in the past, causing all kinds of havoc along the way.

The binary stars of Alpha Centauri show us that even stars can have eccentric orbits around one-another. Proxima Centauri shows us that a star can also have a planet-like orbit. An orbit is an *environmental* characteristic for an object, not a good differentiator for the type of object in that environment. No biologist would say that a marine iguana is not an iguana just because it lives in the ocean. It's still a type of iguana.

Other Orbital Characteristics to Consider

Inclination: One of the older arguments for removing Pluto from the list of planets was because its orbit crosses Neptune's orbit. But in reality, their paths never cross because of Pluto's orbital inclination.

When drawn on a flat piece of paper, the orbits appear to cross, but the Solar System is three dimensional. Look at it edge-on and those intersection points don't actually intersect. Pluto's orbit is higher than Neptune's at those points.

Every orbit is tilted relative to something else. We use three different methods to measure the inclination, or tilt, of an orbit. The first, original method was relative to the “plane of the ecliptic”, which is just a fancy way of saying from the perspective of Earth’s orbit. But Earth's orbit is not un-tilted itself. The second method is to measure it from the Sun’s equator. But what if the Sun’s axis of rotation is tilted, like Earth’s is?

In fact, it is. The Sun’s tilt is one of the clues hinting at the presence of a large planet deep in the Kuiper belt, far beyond where our current telescopes can see.

The third method is to compare the inclination of an orbit to the “invariable plane.” To oversimplify, this is basically an average of the angular momentum of all the mass in the Solar System.

There’s also a tilt relative to the galaxy’s “invariable plane”, but no one seems to use that one. As you might guess, all these values are different and useful for different purposes. None of them are useful for defining a planet.

Eccentricity: This tells us how round, or not round, an orbit is. The more eccentric an orbit is, the greater the difference is between how close and how far away an object travels from its orbital anchor.

Some astronomers have used eccentricity as a way to differentiate between asteroids and comets. But stretched orbits are not unique to comets. Anything can be in an eccentric orbit. There are stars in extremely eccentric orbits around the black hole at the center of our galaxy.

Apsis: The eccentricity is just a ratio. We also need to know the actual values for how close, how far, and how wide an orbit goes. These are actual distances known as the periapsis, apoapsis and semi-major axis. Two of those terms are specialized depending on what anchors the orbit. If the anchor is the Sun, they’re called perihelion and aphelion. If it’s the Earth, they are perigee and apogee. For Jupiter, it’s perijove and apojove. for a generic star, it’s periastron and apoastron… etc.

Other commonly used orbital parameters include: Mean Anomaly, Longitude of Ascending Node, Argument of Perihelion, Time of Perihelion, Average Orbital Speed, etc…

There are a lot of different qualities an orbit can have, but the only ones being used to determine planethood is "around the Sun," which is not unique to planets. It includes just about everything in the Solar System that isn't a moon. And "cleared the neighborhood," which is basically an attempt to set a lower limit on the mass for a planet without actually picking a specific mass. Unfortunately, clearing the neighborhood can also be dependent on other things in other orbits. Ceres could have cleared its neighborhood if not for Jupiter.

Orbits can be more complicated and variable than the objects that occupy them. If you look hard enough, you can find references for "box orbits" (found in tri-axial systems that do not possess a symmetry about any of it axes) and "loop orbits" (found in spherically symmetric or axisymmetric systems), among others. Some references consider parabolic and hyperbolic paths to be orbits, others do not. There is a time component to orbits—they can be short-term stable but long-term unstable. This is related to something called the Secular Phenomena of an orbit. Earth has a secular phenomenon known as the Milankovitch cycle. It causes our axial tilt to precess every few thousand years.

We're not even going to start on the topics of relativistic curvatures of spacetime and the geodesics of Riemannian manifolds. Or the classic three-body problem. Orbits of multi-star systems can get extremely complicated very quickly. Figure-eight orbits have been proven to be possible too.

And, orbits can *osculate*. The perturbations of passing stars and other planet-like things can change the dynamics of any orbit at any time.

Should this really be part of how an object is classified?

Section Three

Ideas for Moving Forward

Chapter Twenty-Two: First Thoughts

How Biology Does Taxonomy:

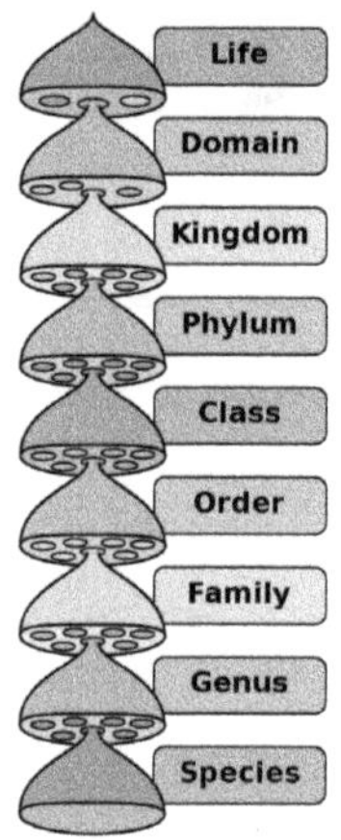

On the left is the grand idea. This taxonomy was created by Carl Linnaeus in 1753, more than 100 years before Charles Darwin published his book, *On The Origin of Species*, in 1859. Linnaeus didn't have the theory of evolution to help him, but he could still see the family resemblance in different species. He was able to make a functional system from his observations. It may have even helped inspire Darwin in his revolutionary work. The diagram below is the tree of life generated by a modern version of the Linnaean system. It's clean, it's simple, and it works.

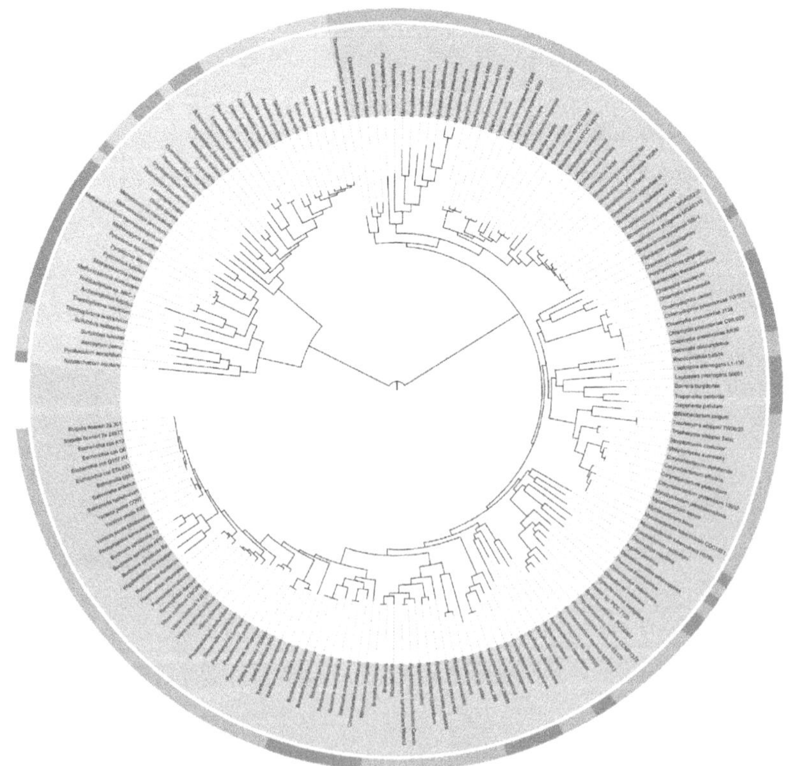

How Astronomy Does Taxonomy:

Now look at this popular Euler diagram of Solar System Bodies. The entire astronomical community should be embarrassed by this. It is chaotic and disorganized. And it leaves out as many categories as it includes. Imagine if it included the cubewanos, the sednoids, the trojans, the scattered disc objects, the various classes of near-Earth asteroids (Apollo, Aten, and Amor), etc?

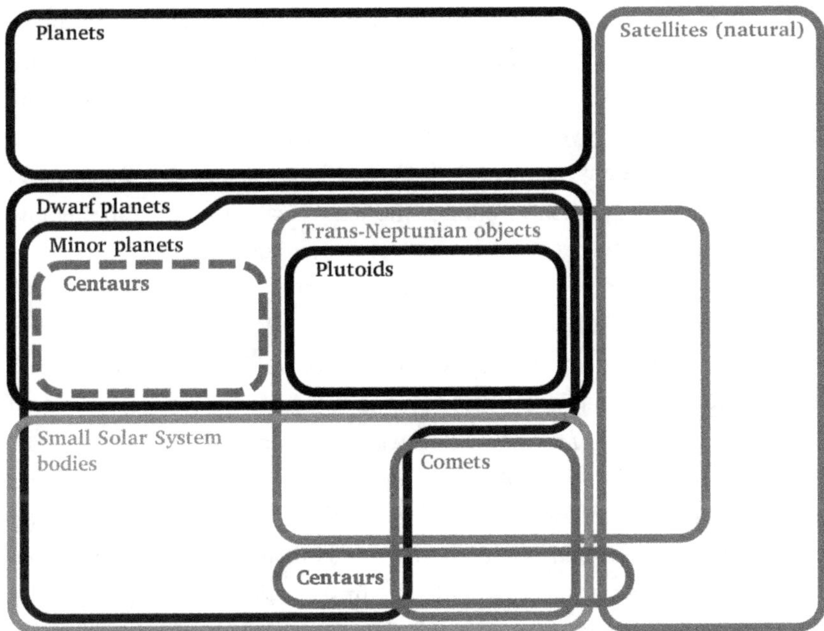

— Source: Tahc / Ariel Provost / SounderBruce

"A library without an index becomes less informative the more it grows"
— Jorge Luis Borges.

Good classification systems are important. They make information useful. Bad classification systems just get in the way and create confusion.

If the IAU wanted to limit the classification of planet to the eight largest things in orbit around the Sun, they could have defined a planet as: "One of the eight largest things in orbit around the Sun". Or better yet, to avoid issues around the definition of an orbit (remember that Jupiter and the

Sun co-orbit an external barycenter), they could have followed the example of the dictionary definition for continents and say, “A planet is any celestial object from the list of Mercury, Venus, Earth, Mars, Jupiter, Saturn, Uranus or Neptune.”

Make no mistake, that was the goal of their definition. But they couldn’t be that blatant and get away with it. So they did it by couching the list in a pseudo-scientific filter that is useless for anything other than forming that exact list.

A filter-based criteria masquerading as a definition can also let things through that were not intended, like our moon, Luna. And it eliminates things that should be included, like all the exoplanets in the galaxy and beyond. The criteria given by the IAU for a planet does nothing to advance the science of astrophysics, and does quite a lot to damage it. We need an entirely different approach to our cosmic classification scheme, preferably one that separates objects and orbits into two separate descriptors.

In the pages ahead we’ll explore some ideas. We’ll begin by analyzing the objects independent of their orbits. Then we’ll consider orbits and orbital influences. Finally we’ll see if we can find a way to incorporate both into a modern system of taxonomy for celestial objects.

Keep in mind — none of this is intended to be definitive. I make no claims to have the authority or the credentials to do that. This is nothing more than a conversation starter. Better minds than mine can take it from there. If they throw it all away and start fresh, as long as we get the ball rolling toward a system that actually makes sense, I will consider this book a success.

Chapter Twenty-Three: Other Ways To Test for Planet-ness

This chapter is for the teachers, parents, their kids, and anyone else in the world who has an interest in the Solar System, but has not made the commitment to require a scientifically rigorous terminology.

Honestly, a lot of professional astronomers and scientists don't use scientifically rigorous terminology in their conversations either. There is nothing wrong with that.

Most planetary scientists refer to the largest moons of the gas giants as planets. If this bothers you, then don't do that. If it appeals to you, then please do call them planets. You have the power and the right to participate in shaping our common language too. This is no different than calling Europe a continent. Not everyone does it. Not everyone has to. Exactly zero crimes have been committed either way.

Most importantly, remember that words are only a method of sharing ideas between people. As long as the people you're talking to understand what's being communicated, mission accomplished.

Okay, but what is a planet, really?

Rather than promulgate a tediously specific list of speciously spurious qualifications, let's do it by examples. The lists on the following pages are compiled in the same sense that the continents, oceans, mountains and streams are. These are places that intelligent and genuinely sincere people have referred to as planets in a public record. These could be newspapers, magazines, scientific papers, on-line forums, legal proclamations, books, or verbal conversations in recorded interviews.

Each list is based on different reasonings. If there is anything in these lists that you personally feel doesn't belong, then leave it out. If there's something you think is missing, add it to your repertoire of planets when you talk or write on the subject. Eventually it will all settle out.

The "Star Trek" Test

If your spaceship is approaching a large round object that it can go into orbit around, and send a landing party to explore (with or without specialized environmental suits), then it's probably a planet. Like mountains, these objects stand apart from the local terrain in a significant way. They are the most dominant things within their orbits.

By this reckoning there are:

Five inner planets: *Mercury, Venus, Earth, Mars,* and *Ceres*.

This should be obvious. Since Ceres is by far the largest thing in the asteroid belt, containing one-third of the total mass of the entire belt, and is the one and only spherical object in that zone, and has an active geology, it would definitely stand-out as a planet from the viewpoint of my spaceship as I pull into orbit around it.

Plus **Fifteen** satellite planets: *Io, Europa, Ganymede, Callisto* (at Jupiter); *Tethys, Dione, Rhea, Titan, Iapetus* (at Saturn); *Titania, Oberon, Umbriel, Ariel* (at Uranus); and *Triton* (at Neptune). Rhea is especially interesting. It is the only planet in the Solar System besides Earth to have an oxygen-rich atmosphere. It's too thin and cold to breathe, but it could be useful to explorers with the right kinds of equipment.

Add **Three** trans-Neptunian bodies: *Pluto, Persephone* (also known as *Eris*), and *Sedna*. I'm including these because, like Ceres, they are the largest and most dominant objects in each of their individual zones (see chart on facing page). Pluto is in the group of Resonant Objects, Eris is in the Extended Scattered Disk, and Sedna is in the Extreme Detached Disk.

This makes a total of **Twenty-Three Planets** in our Solar System.

I excluded the gas giants because they have no surface geology to land on. They are better described as an intermediate phase between 'planet' and 'brown dwarf'. To my sensibilities, these are not planets. They are gas balls.

Also, Luna and Charon are excluded in the same sense that Makemake and Haumea are excluded. They are not the obviously dominant objects within their orbits.

Remember, this is not a hard rule. It's just a thought experiment to get us warmed-up. Apply your own ideas to this list and let the debate begin.

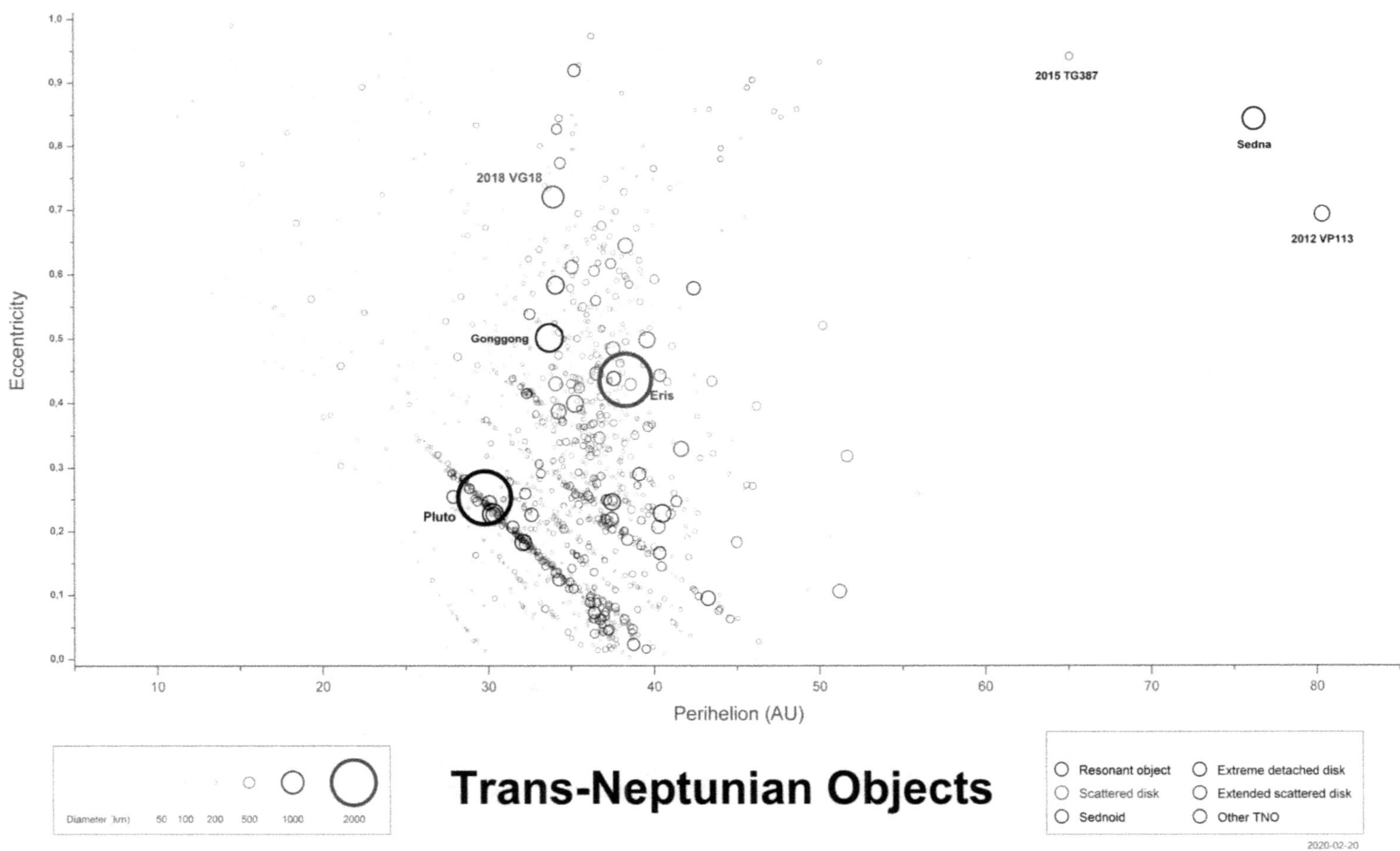
Eccentricity
1,0
0,9
0,8
0,7
0,6
0,5
0,4
0,3
0,2
0,1
0,0
2015 TG387
Sedna
2012 VP113
2018 VG18
Gonggong
Eris
Pluto
10
20
30
40
50
60
70
80
Perihelion (AU)
Diameter (km)
50
100
200
500
1000
2000
Trans-Neptunian Objects
Resonant object
Scattered disk
Sednoid
Extreme detached disk
Extended scattered disk
Other TNO
2020-02-20
Daniel Bamberger

The Roundness Test

One of the favored determinants among planetary scientists is the idea that being spherical is good enough. I don't like this one because ice and water become spherical at much smaller masses than rock or iron.

Having said that, the planetologists have good reason for preferring it. Becoming spherical is an important turning point in the evolution of a planet. When gravity squeezes it into a ball, things start to move around. Stuff squirts through the cracks and pores (volcanism). Tectonic activity begins and the planet becomes geologically "alive."

But there's a problem with ice. If the temperature is cold enough, ice can remain porous at sizes up to 1000 km in diameter. In other words, it is somewhat "fluffy". It can be spherical, but the gravity of these low-density, low-mass objects is not enough to crush it together and start the geological engine. Rhea might fall into this category, for example. We won't know until we go there and learn more about it.

Rock gives-in to gravity somewhere around 900 km, as demonstrated by Ceres. Since Ceres is the smallest confirmed rocky sphere in the solar system, it is a logical choice for a minimum size for a planet. But let's be generous and use the slightly smaller value of 800 km for the cut-off size, just in case we missed something somewhere. This will only add two more known objects to the list and gives us a total of **Thirty Planets**.

These are the **Thirty Planets** in order from largest to smallest:

Earth, Venus, Mars, Ganymede$_{(J)}$, Titan$_{(S)}$, Mercury, Callisto$_{(J)}$, Io$_{(J)}$, Luna, Europa$_{(J)}$, Triton$_{(N)}$, Pluto, Persephone (aka Eris), Haumea, Titania$_{(U)}$, Rhea$_{(S)}$, Oberon$_{(U)}$, Iapetus$_{(S)}$, Makemake, Gonggong, Charon$_{(P)}$, Umbriel$_{(U)}$, Ariel$_{(U)}$, Dione$_{(S)}$, Quaoar, Tethys$_{(S)}$, Sedna, Ceres, Orcus, Salacia.

Natural satellites of larger objects are denoted with a subscript; J=Jupiter, S=Saturn, U=Uranus, N=Neptune, P=Pluto. Since Luna is gravitationally bound to the Sun with twice the force as it is to Earth, its satellite status is disputed and not denoted. Giant gas balls are excluded due to a lack of a discrete surface to measure.

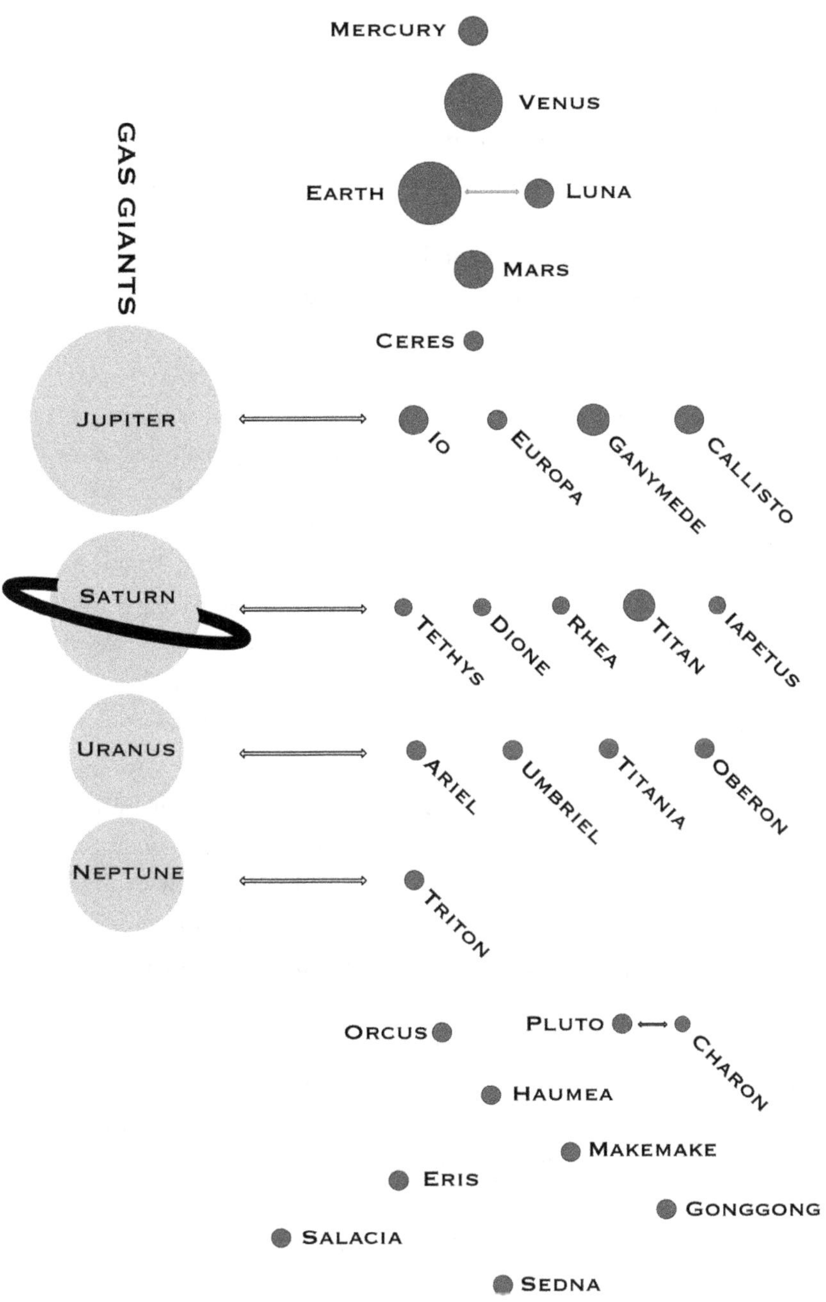

Diagram of the 30 planets (not to scale)

The Usability Test

If we ignore size and shape and consider surface gravity, we can get an idea for what doing work on a planet might be like. This is not intended to be a complicated, rigorous formula. It's just the idea that a person should be able to walk around and do meaningful work without the risk of being thrown into orbit from a minor industrial accident. Obviously, this requires a surface to stand on, so once again, the giant gas balls are excluded.

The order of the planets will be different from lists based on mass or size. Size and gravity will vary depending on the density of whatever the planet is made of.

The Earth's surface gravity is 9.8 m/s^2 Let's round this up to 10 m/s^2 for the sake of convenience, then take five percent of that as a "usability" cutoff. Why five percent? It's just a starting point. It can be adjusted up or down if experiments determine a better ratio. To do that, we may have to actually go visit some of these planets and do some work there.

At five percent of Earth's gravity, a 200 lb. man will weigh 10 lbs. As he walks, it will be easy for him to pick up a lot of speed. Here's the danger — he still has 200 lbs. of *momentum*, but only five percent of the *friction.* Trying to stop or change direction will be a lot like walking on ice.

Five percent of 10 m/s^2 is 0.5 m/s^2. Only **Thirteen Planets** have a surface gravity greater than 0.5 m/s^2. For some people, the call of adventure is more important than safety or convenience. They may be willing to take the risk and put forth the effort to live and work on a planet with less gravity. For them, we've included twelve additional mesoplanets to explore.

The chart on the facing page shows the planet's name, surface gravity in meters per second squared (**g in m/s²**), the weight of a 200 lb man on that planet, in pounds (**200 lb. =**), the escape velocity of the planet in kilometers per hour (**V_e km/h**), and the absolute minimum amount of TNT required, in pounds measured on Earth, to launch that man into orbit (**lb. TNT**).

Note: Sedna probably falls somewhere between Charon and Iapetus, but we don't have accurate enough data to be sure, so I left it off the list.

(Apologies for the mix of SI and Imperial units. 100 kg is 100 kg whether it's here or on the moon. But lbs of force are more intuitive for comparing weight.)

	Name	g in m/s^2	200 lb. =	V_e km/h	lb. TNT
	Earth	9.8	200	40260	2720
	Venus	8.87	181	37290	2340
	Mars	3.72	75.8	18072	548
	Mercury	3.68	75.2	15300	392
	Io	1.79	36.6	9216	143
	Luna	1.62	33.0	8568	123
	Ganymede	1.42	29.0	9867	163
	Titan	1.36	27.8	9504	152
	Europa	1.314	26.8	7290	89
	Callisto	1.236	25.2	8784	130
	Eris	0.819	16.7	4932	41
Planets	Triton	0.779	15.9	5256	47
	Pluto	0.650	13.26	4428	33
	Haumea	0.430	8.76	3024	15.5
	Makemake	0.404	8.24	2880	14.0
	Titania	0.364	7.42	2772	12.9
	Oberon	0.354	7.22	2628	11.6
	Quaoar	0.298	6.08	2077	7.3
	Charon	0.288	5.87	2088	7.3
	Ceres	0.284	5.79	1836	5.7
	Rhea	0.270	5.51	2286	8.8
	Umbriel	0.250	5.10	1872	5.9
Mesoplanets	Ariel	0.249	5.08	2016	6.8
	Dione	0.231	4.71	1836	5.7
	Iapetus	0.223	4.55	2052	7.1

The Data Science Test

Data scientists like to look for natural breaks in the data. If there are groupings around specific values, or if a chart suddenly changes its slope at a particular point, it usually means something important is happening there.

The chart on the facing page lists the 45 largest objects in the Solar System, sorted according to size in kilometers. Again, the giant gas balls are omitted. At this scale, their bars won't fit on the page—not even if we double the full length of the page. They are absurdly large.

What we're looking for are natural breaks in the data. As you can see, there are four places where the shape of the graph makes a dramatic change.

A: Earth and Venus are in a class by themselves. Mars is a distant third. From this data alone, it could be argued that there are only two planets in our Solar System: Earth and Venus.

B: Here we have four objects in a fairly linear grouping. Ganymede, Callisto and Titan are clearly in the same family as Mercury.

C: Another almost-linear grouping, but with with a slightly sharper slope. Notice how the slope eases at Io and Eris.

D: After Eris, the graph changes its character. Other than a single small drop between Makemake and Haumea, all the drama is over. It's a slow, gradual decline from here. This is known as the "long tail" of the data. The body of the beast ends where the tail begins. Eris is the smallest object in the body of the beast.

Graphs that make a dramatic change in slope like this are commonly called "hockey stick" graphs. They can be found in climate science, population growth, stress fractures and more. The bend is caused by a tipping point being crossed. The elbow of the hockey stick is where the critical change happens.

Based on this information, it looks like **Thirteen Planets** stand apart from the rest. The most significant break-out occurs at planet Persephone (aka Eris). Draw your own conclusions, but I would say that everything from Eris (Persephone) to Earth is in the planet region of the graph.

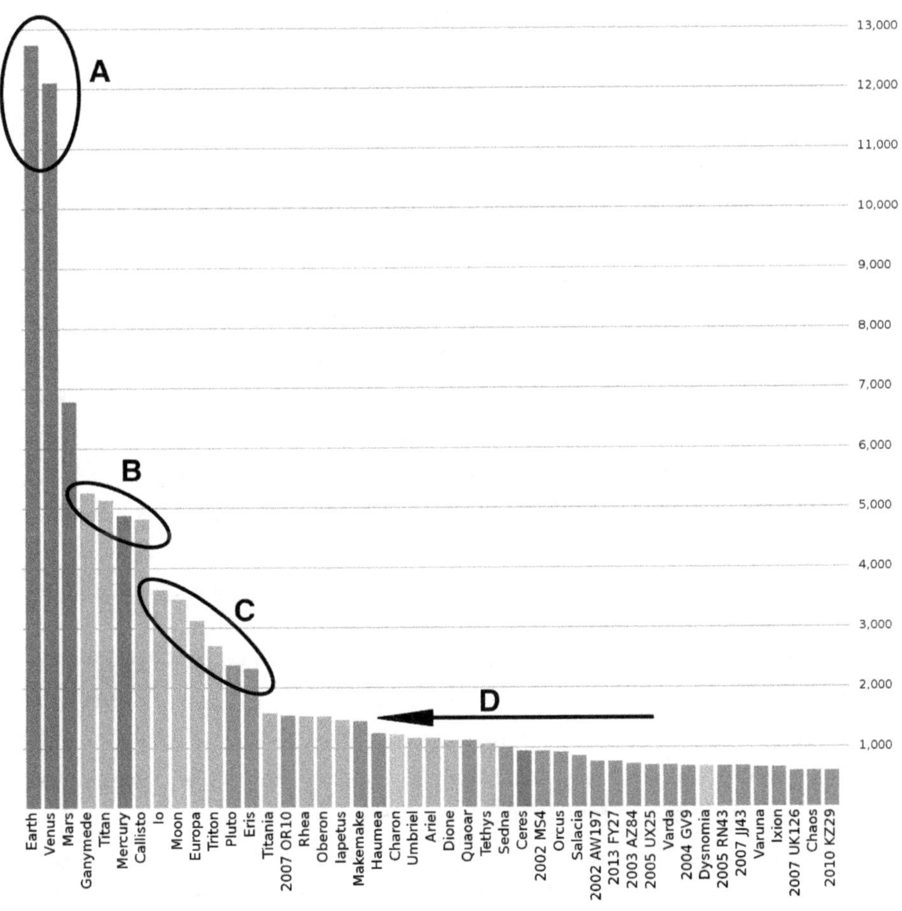

Illustration credit: Tbayboy (modified)

Observations and Conclusions

As anyone can see, there are a multitude of ways to judge the planet-ness of a celestial body. The previous eight pages are just as scientific as anything the IAU offered regarding the word planet. Maybe more so.

All of this is more evidence that the word "planet" is not suitable as a rigorously defined scientific concept. The IAU has not improved our understanding of the universe or our reality. By suggesting that Jupiter and Earth have more in common than Ceres and Mars do, they have made our understanding worse.

Metaphorically speaking, a house is still a house whether it is alone on a tiny island, or in a continental countryside surrounded by cornfields, or deep in the inner city of a large metroplex sitting in the shadow of a high-rise corporate office building.

It is also true that the lifestyles of the people in those houses will be different. Likewise, the conditions of a planet will be profoundly different depending on its orbit. Orbits are important, but they are a *separate* thing. They are the environment. The environment is not the animal.

So, where does this leave us with the word planet? My suggestion would be for the IAU to do what the USGS did with the words mountain and lake; issue a formal statement that there is no scientific definition for these common words. Then move on to creating a new and *useful* taxonomy of cosmological bodies that is orderly and logically consistent.

Planets, like biological organisms, can grow and evolve over time. Rather than splitting off to form new species, celestial objects can collide and bond together forming a larger object. Or, if the kinetic energy is high and gravities are low, they crash, smash, and spew chunks of debris across the cosmos for other growing planets to feed upon.

This process is loosely analogous to evolution and extinction. Along the way these objects are shaped by their environments, just like organisms are. We can use this in our effort to create a new taxonomy.

Chapter Twenty-Four: Planetesimals, Planetoids, Minor Planets, Mesoplanets, Dwarf Planets, Planetars, and Other Things Not Generally Regarded As Planets In Spite Of Their Names. Also Planets.

Planet; a tortured and ridiculously over-burdened word. It's a thing, it's an orbit, it's the quality of a neighborhood, it's a prefix and a qualifier and a category. It's the points of light in the sky that move. It's where we send the away teams and robotic probes.

If you do a google image search for "planet", you won't see orbits or neighborhoods or mass equations. You'll see spherical objects—great balls of rock and/or ice. Sometimes enveloped in a gaseous atmosphere. Sometimes harboring pools of liquid, sometimes barren and exposed to the vacuum of space.

But if you remove the rocky orbs from your search results and replace them with maps of orbits, equations for mass, and overviews of vast regions of the Solar System, scant few people will understand the connection.

Since comets, meteors and asteroids are all just asteroids in transient situations, new words were spawned to try and differentiate them. The word 'asteroid' was deprecated a long time ago in favor of the term "minor planet." But some astronomers preferred the one-word term 'planetoid' over the two-word 'minor planet'.

There's an obvious size difference between asteroids/planetoids that can be seen in a telescope vs. any meteors witnessed by modern humans, so the term planetesimal was created for the smallest planetoids that we know are out there. But no specific boundary was ever established.

In 2006 the IAU decided to sweep all this away and changed the preferred term to "Small Solar System Bodies" (SSSBs). The only definition given is that they are not planets, they are not dwarf planets, and they are not natural satellites (aka moons).

There is good logic behind this. The variability of types of objects and their paths is far too complicated for simple, individual, one-word descriptors. There are asteroids that have satellites of their own (we know of more than 400 so far). There are also asteroids known as "active asteroids" because a tail, or a coma, is occasionally detectable. But these tend to stay in the vicinity of Jupiter instead of swooping down toward the Sun. Whatever you find out there, it's an SSSB.

> **Nature does not care about the nice tidy categories we impose upon it.**
>
> — Stefan Milo

Unfortunately, astronomers don't seem to care much either. In addition to SSSBs, the IAU still actively maintains the term Comet. There are SSSBs that are explicitly comets, like 11801 LINEAR, or 2060 Chiron (which also happens to have a ring system, like Saturn and Uranus). Somehow, Ceres is classified as an active asteroid (comet-like), even though it has no tail.

— https://en.wikipedia.org/wiki/Active_asteroid#Members

Consistency does not seem to be a strength of the IAU. None of the terms Asteroid, Minor Planet, planetoid, or planetesimal were ever used with any consistency.

As far as I can tell, Isaac Asimov was the only person to use the term mesoplanet, meaning "middle-sized planet." It was his version of the modern term 'dwarf planet,' created decades before the IAU issued it. (Honestly, mesoplanet is a better term.)

The IAU was right to deprecate all of them. But lets be honest; "Small Solar System Bodies" is an awkwardly clumsy replacement. We can do better.

Let's take a look at that Euler diagram again:

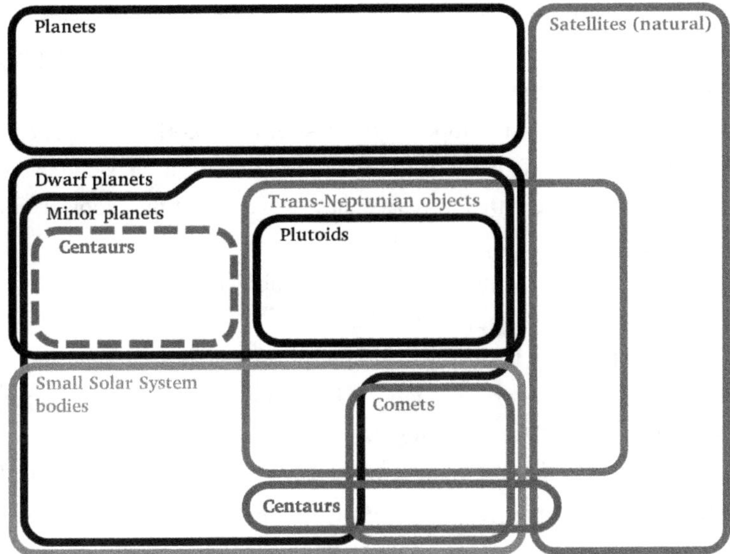

I don't think anyone can call the jumble of categories in astronomy "nice and tidy." There's no mystery why these categories are routinely ignored.

I'm not going to pretend that I have the expertise or credentials to fix this singlehandedly. But I would love to participate in a room of qualified people who are honestly dedicated to solving the problem.

Regarding the status of Pluto, there is only one obvious solution to this conundrum. We must divorce the word 'planet' from any hard, specific scientific meaning. The word 'planet' belongs more to the realm of literature than to the natural sciences.

'Planet' and 'moon' are common words, used commonly. Just like mountain and river and continent and ocean and crab and dog and cat. Is a mountain lion a lion, or a cat, or both, or neither? It all depends on the context and what's being communicated. The biologists of the world have their own unique taxonomy to serve their more technical purposes.

Speaking of technical purposes, what do Jupiter and Mercury have in common other than the fact that they both orbit the same Sun? How is this scientifically useful?

We should give up the obvious falsehood that the four gas giants belong in the same category of objects as the four terrestrial planets. Literally everything about them is different, often by orders of magnitude—Composition, density, scale, topology, celestial neighborhoods… These are clearly two separate categories of things. Putting them in the same category is just bad science. The only similarity they share is that they look like stars when viewed from Earth with naked eyes.

If the gas giants are not true planetary objects in the same sense as Mercury or Earth, what should we call them?

The word planetar is not new. It is a portmanteau of *planet* and *star.* Planetar was one of the original names for brown dwarfs, but is no longer used in that context. Since Jupiter is a sub-brown dwarf in a planetary setting, and since calling it an asteroid (meaning star-like) would just be confusing to most people (even though it's technically more accurate), the word *planetar* is the most appropriate choice we have available for this kind of object.

If we adopt this concept for all the gas giants, then we'll lose four planets but gain four new planetars! According to the logic of Neil deGrasse Tyson (who said Pluto should be happy as the king of the dwarf planets), Jupiter and the others should be happy with their upgrade to planetar.

If we consider Jupiter and Saturn, et al, to be "star-like", then their "moons" become "planet-like". It might be tempting to call them *planetoids*, But this is problematic since planetoid is already a co-synonym with asteroid, minor planet, and Small Solar System Body.

There's really no need for a different category anyway. The word Planet describes them well enough. If we need to differentiate them, we can add the qualifier *satellite*. Io, Europa, Ganymede, Callisto, titan, et al, will be the satellite planets. If that's too troubling for most people, then the term mesoplanet is still available.

Regarding the objects known as Ceres, Pluto, Eris and Sedna, it doesn't matter that there are other things near their orbits. Where they are doesn't change what they are. People who are more concerned with the 'where' than the 'what' should create a new name for that relationship. The word planet has never carried that kind of meaning.

Someone may be thinking that this is why the asteroids were not considered planets — because of where they are. But that's not accurate.

The asteroids were a mystery. Astronomers didn't know what they were. By the late 1800s it was decided they were "minor planets." Notice the word *planet*. At the time, the designation of minor planet was a reference to their size, not the quality of their neighborhood.

And we don't need to create an arbitrary filter to determine which ones to put on a list and which ones to leave off. We can let people figure it out for themselves. Trust me, this will work.

Consider: Most people have heard of the Himalayas. And most people have heard of the tallest mountain in the Himalayas, Mt. Everest. What's the second highest mountain there? Some sources report that it's K2, but K2 is not a Himalayan mountain. It's in the neighboring mountain range, known as the Karakoram range. The second highest mountain in the Himalayas is Kanchengjunga.

Notice the pattern. People remember the biggest things, even when they're in different areas. People rarely consider the second in line to be as memorable. How many people know the third or fourth highest mountains in the Himalayas? (in order, they are: Lhotse, Makalu, Cho Oyo, etc…)

All these "lesser" mountains are much taller than Mt. Fuji in Japan, or Mount Olympus in Greece, or Mont Blanc on the border between Switzerland, France, and Italy. America has Denali in Alaska, Mt. Hood in Washington, Mauna Loa in Hawaii, and controversially, Mount Rushmore in South Dakota.

These are memorable and noteworthy because they are the most significant objects in their region. They are not all the highest, They are not all formed the same way. Some are not even true mountains. We call them that because they are the dominant feature in their local environments.

Every state has a largest city, and every state has a state capitol. Sometimes they are the same, sometimes not. Some states (RI, DE) are smaller than counties in other states. San Bernardino county in California has more than ten times the land area of the state of Delaware. Cook County in Illinois has almost ten times the population of the entire state of Wyoming. Does this invalidate Dover, DE or Cheyenne, WY as state capitals, even though other non-capital cities govern more people and more area?

We can treat planets in a similar way. Ceres is the most significant object in its region. Ditto with Pluto. Ditto with Eris. And with all the classic planets as well. There will always be one largest thing in an orbit. If we grant the title of Planet to the most significant thing—one planet per orbit—this accomplishes two things:

1. It solves the issue of clearing the neighborhood. There can be a million other things in the vicinity, but only one is the planet of that region.

2. It solves the problem of planet-sized moons. Luna is smaller than Earth, so it's not the planet in their shared orbit. Same with Charon.

Now we have a much more interesting (and inspiring) Solar System. It is important to remember that this is not a scientific designation. This is only intended for common usage by ordinary people to highlight the most interesting places in the Solar System.

Alternately, in keeping with the way most planetologists use the word, planets can be described as any celestial object that is between the size of Ceres (the smallest size where Rock becomes a sphere) up to the size of the smallest star-like things. Whether that's a red dwarf, or a brown dwarf, or a gas giant is debatable. In my opinion, the gas giants are more star-like than planet-like. But you should make that determination for yourself.

And a planet can be in any orbit — around a star, around a gas giant, around another planet or even in a Lissajou orbit around a Lagrangian point, if we ever find one in that situation.

The fact that Pluto's and Ceres's neighborhoods are not "cleared" can easily be handled with an additional qualifier. They are planets in *cluttered* solar orbits. But the word cluttered is subjective. Remember than Neptune and Pluto share part of their orbits, and Jupiter's orbit is polluted with debris. Saying these are planets in a cluttered orbit is just as valid as saying they are dwarf planets. Except the term "cluttered orbit" descriptor is more obvious and more useful.

Chapter Twenty-Five: Dangerous Ideas For An Ambitious Astrophysicist Looking For Trouble

If you want to leave a mark, you've got to make some waves, and if you want to make waves, you've got to rock the boat. Just make sure your boat doesn't capsize and the waves you create do more to clear the water and and wash away debris rather than cause chaos and destruction.

Here are a few ideas. They are only ideas. I don't want to mislead anyone into thinking that I have any special expertise over these things. They are offered for whatever use they may be to anyone for any purpose, whether a new space mission, a chapter or two in a science fiction novel, or just a good laugh.

I'm old enough to remember the Apollo program first hand. July 20, 1969—my heart pounded in my chest as an eight year old kid. I was glued to the television watching the LEM hover over the Lunar surface, fuel running low, searching for a safe spot to land. Then the heart-stopping landing, followed by the raw elation of knowing that I had seen real history being made. It was a special moment that would be remembered and talked about for a thousand years to come. And I got to see it on live TV.

I barely slept that night. I remember waking up especially early, running to the television for any updates, the agonizing anticipation until finally, Neil Armstrong and Buzz Aldrin exited the LEM and took those first steps on the lunar surface. It was a new world, both in space and on Earth.

Less than a decade later Carl Sagan was famous for his role in the Viking missions to Mars, and then the Voyager missions to the survey the outer planets. Most of what we know about Uranus and Neptune came from those Voyager missions.

Those were exciting times. They inspired me to get serious about math and pursue an education in engineering. I know I'm not the only one. We didn't all get to work on the space programs, but that's okay. The world needs good engineers and scientists everywhere. We should inspire more people to pursue those paths. STEM programs in high schools are great, but the funnel into those programs starts much earlier.

There is a lot to learn out there. I'm sure a lot of it can help us achieve a better world here at home too. New frontiers and new opportunities are waiting. All we need to do is extend our reach. Someone has to be the driver to make that happen. It won't be easy. These are the kinds of things that take whole careers to achieve.

First Idea: We need to know a lot more about objects in the asteroid belt than just size, mass and reflectivity. All of our Small Solar System Bodies (or space rubble, per NASA), might be better sorted according to other more interesting qualities than the current sorting scheme based on the order of discovery, or alphabetical by name.

— https://minorplanetcenter.net/iau/lists/MPNames.html — The IAU maintains a significant database of orbital data on the asteroids too, which you can find under the "data" tab at https://minorplanetcenter.net

Let's send a fleet of small autonomous survey probes into the main belt, possibly propelled via solar sails or solar powered ion drives (or something else), and have them circumnavigate the belt from within it.

A second fleet of mini-probes should land on selected asteroids known to be stable and in low eccentricity orbits. One hundred and eighty of these mini-probes could be established every two degrees around the circle, or send more probes and put them closer together if possible.

A third fleet can be established on a handful of asteroids from the Hilda group. The Hildas have the unique quality of darting in and out of the main belt as they pinball between Jupiter's L4, L3 and L5 positions.

All these probes should be designed to learn as much as possible about the stability of various parts of the belt. We don't know much about the asteroids smaller than one kilometer in size. They're pretty hard to see from Earth. How often to they collide? How solid or loose are they? How hostile is that environment? How easy or hard will it be to mine these structures or build on them using autonomous machines?

Of course they will transmit photos and a whole suite of data to us about everything they find. This will really flesh-out our database of information about the smallest and most populous parts of our Solar System. Then maybe we can have enough data to create a more logically consistent taxonomic scheme.

Why do this? For the same reason people have always surveyed the wilderness of the world. There is a whole lot of real estate out there waiting to be exploited. We won't know what we can do with it until we know what we're working with. Asteroid mining is only one potential use for this resource. There are others…

Second Idea: Now that we've got the infield covered, we need to learn more about the outfield. The outer regions of the Solar System are sure to be endowed with different landscapes and different riches to be found.

Don't make assumptions. I'm not talking about the personal wealth kind of riches. I'm talking about whole new economies. Michael Faraday's explorations of electro-magnetism didn't have much practical use for many decades. But today, 99% of our global economy is utterly dependent on the groundwork he laid. No one could have foreseen that at the time.

Exploring the trans-neptunian region will be a bit more difficult. We need another fleet of small survey craft to penetrate the regions beyond Pluto and see what's out there in an up-close and personal way.

Idea Three: By now we've got a whole bunch of devices scattered far and wide across the Solar System. They're going to need to get their data back to Earth somehow.

Having all these transmitters track the Earth, and having various receiving stations on Earth track all these things, is a lot more complicated than it needs to be. Fortunately, we've already got the basic infrastructure in place for a celestial communications network.

We can take advantage of the Solar System being mostly disk shaped. Those probes we put on low-eccentricity asteroids spaced every degree or two around the belt can be the "cell towers" for an interplanetary cellular network. Anything outside the belt can simply transmit toward the Sun. Whatever cell is in line between the Sun and the signal can intercept it and shunt it toward the next cells around the circle.

Whichever cell is in line with the Earth and Sun will send the signal to Earth. Since this will be an open and publicly accessible technology, any citizen scientist who wants to play with the data can point their receivers away from the Sun to find the signal.

Anything inside the belt (Mars, Mercury, Venus, others) can send their signals away from the direction of the Sun, and also listen for signals from each other coming from that same direction, regardless of their actual locations. This makes it trivial to communicate between planets when they're on opposite sides of the Sun.

Now, this is a grossly oversimplified description. We're talking about an hour and a half light-speed trip around the belt that will have to be accounted for. In practice, the transmitters on the asteroidal cells will need to send their signals inward toward the Sun, but also tangent to the belt. Receivers outside the belt can't point directly toward the Sun for a signal. There's too much background noise. They'll have to aim at the edges of the belt to get those tangential signals.

Idea Four: Now that we've got a solid base of operations, a reliable communications infrastructure, and a little experience under our "belt", it's time to start construction on something more serious.

In modern astronomy, one of the most useful tools we have is something called parallax. It's not a technology. It's the practice of taking a photo of the sky, waiting six months for the Earth to go halfway around the Sun, then take another photo from there. It's the astronomical equivalent of having two eyeballs. The parallax shift gives us 3-D stereoscopic vision. We can see depth and get accurate distances to the stars this way.

Another technique is to coordinate multiple radio telescopes from around the world to look in the same part of the sky at the same time, then blend those signals into a single, "virtual" dish the size of the entire Earth.

Now—imagine a virtual radio dish the size of the entire asteroid belt. Imagine the parallax images from there. And, you don't have to wait six months. We can have a ring of telescopes six AU wide—a giant compound eye to peer deeper into the cosmos than ever possible before.

Idea Five: Since all the planets of our system have different lengths of orbits, different days, different gravity wells, and move at different velocities, getting our clocks synchronized is going to be a challenge. We might as well work out a system of interstellar time keeping.

The idea of a "stardate" was introduced in the TV show Star Trek to avoid the problem of relativistic time dilation due to faster-than-light travel. But they never explained how it works. Then they completely ignored time dilation anyway. Such is life in a TV show.

Here's a thought experiment— imagine you're in suspended animation on an autonomous space ship traveling from Earth to Mars. On the way, something goes wrong. Your ship wakes you up on Saturn's moon Rhea instead. Fortunately, you've got all the technology needed to survive and build your habitat there. But how much time has passed? What year is it?

You radio Earth via the asteroidal cell network, but it will take hours to get a response. In the mean time, you can use your knowledge of the Solar System to get an answer.

Pluto's orbit is 248 years. Jupiter's orbit is 11.8 years. Jupiter goes around the Sun 21 times for every one time Pluto does. We can use their relative positions like hands on a clock to find the current year. The hour, minute and second hands on this clock are analogous to Pluto's, Jupiter's and Earth's orbital positions. Except, you don't know what year it is, so you don't know where to look for any of them. You'll have to scan all 360 degrees of sky. it would be a lot easier if you were outside of the Solar System, someplace where you can see the whole system at once.

Fortunately, at the center of our galaxy is a black hole surrounded by stars in highly eccentric orbits with periods ranging from 9.9 years to 166 years (see the graphic illustration on next page).

Of these, S14 has the highest eccentricity. It's almost a straight line moving back and forth in a 55.3 year cycle. S8 has the second highest eccentricity in a long oval going back and forth every 92.9 years. Three of these stars go clockwise (S1, S2, S14, shown in grey) and three go counterclockwise (S8, S12, S13, shown in black). An accurate reading of their relative positions can give us a current date.

All their orbits precess in a rosette formation (apsidal precession). This precession can be used to determine timeframes in the thousands of years, just like the hands on a clock.

By adding an offset equal to the number of light-years from the galactic center to any location, a common "stardate" can be calculated. And, this clock is visible via radio telescopes from anywhere in the galaxy.

The "Hands" of Our Galactic Clock:

Right Ascension difference from 17h 45m 40.045s

+0.5″ +0.4″ +0.3″ +0.2″ +0.1″ 0.0″ −0.1″ −0.2″

Declination difference from −29° 0′ 27.9″

+0.5″ +0.4″ +0.3″ +0.2″ +0.1″ 0.0″ −0.1″ −0.2″ −0.3″ −0.4″ −0.5″

S12 S1 S2 S14 S13 S8

Orbits of some bodies of the Solar System
(Sedna, Eris, Pluto and Neptune)
at the same scale for comparison

— Image Credit: Cmglee, from data found at https://iopscience.iop.org/article/10.1086/430667/pdf

How apsidal precession works:

Each orbit rotates to a new orientation.

Its path traces a full "spirograph" over thousands of years.

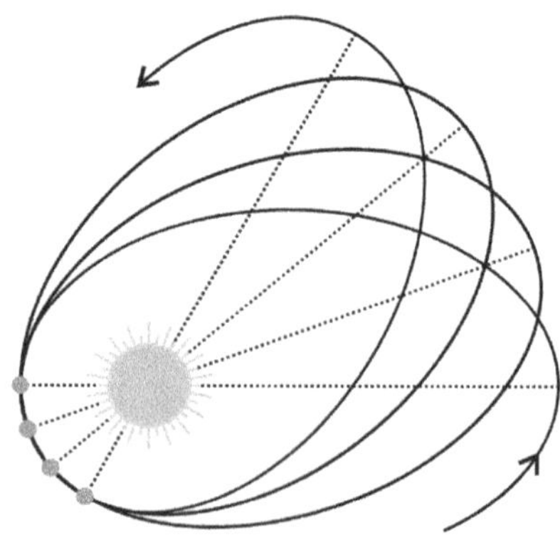

Our Sun orbits the galactic plane every 225 million years. The angle from our Sun, to the galactic center, to the galaxy of Andromeda is a knowable angle. This can be our million-year galactic calendar. Using the direction of Andromeda as a zero-reference position we can calculate a unique time and date with the granularity of days, over at least the next quarter of a billion years.

It should be noted that observing the stars at the center of the galaxy is currently a non-trivial exercise. This could change—*will* change—as our technology advances. It may be a few more centuries before anything like this is truly useful to humanity, if ever. But let's be hopeful.

By then we may have discovered some other fundamental property of time and space that's more convenient to use. One thing is certain—we won't find it if no one is looking for it.

Idea Six: Now that we've got a better view of all the exoplanets and rogue planets and who knows what else we'll discover out there—maybe even a society of friendly civilized aliens—we need to start thinking about interstellar travel. The answer to this is hidden in the unknown properties of Time and Space.

I have a nagging feeling that space-time relativity might be a little bit like the Ptolemaic system. It's super-accurate as far as we are able to test it, but that doesn't mean it has to be the best model, or even the correct model. There might be a better model we just haven't thought of yet. The existence of dark matter and dark energy prove that there are layers of reality we have yet to understand.

A submarine can't travel faster than the speed of sound, but an aircraft can, and there is no speed of sound in space. Maybe we're "underwater" in space-time, like the creatures who exist around thermal vents at the bottom of the ocean, unaware of what lies above. Like them, we just haven't found a way to reach the surface boundary yet.

Idea Seven: Secure your immortality and be the next Carl Linnaeus. Create a taxonomy for celestial objects that is logical, consistent, *useful*, and will endure all the variety of situations and conditions we are likely to encounter over the next few hundred (or thousand) years.

Final Word: How You Can Make Pluto a Planet Again

Don't fall victim to the logical fallacy known as an *appeal to authority.* The IAU's definition does not pass scrutiny in spite of their self-proclaimed authority. Anyone can see this. The emperor has no clothes.

The IAU's definition is only one opinion in this controversy. There are just as many credentialed scientists who disagree with it.

Teach the facts, not the dogma. It is a fact that other trans-neptunian objects have orbits similar to Pluto's. But unlike the other objects out there, Pluto has an atmosphere, changing seasons, polar ice caps, mountains and volcanoes. It has vast open plains and an active geology. According to a majority of scientists, this is without a doubt a planetary body.

There are asteroids with orbits similar to Earth's and Mars's. This does not prove that Earth and Mars cannot be planets. The existence of other trans-neptunian objects does not prove that Pluto isn't a planet.

Whenever you mention the planet Pluto, be sure to call it a *planet.* Every time you use the word planet, your words are a vote for what the word planet shall mean. Your vote counts, and unlike political races, you can vote as often as you like. So start the conversation. Keep it going.

Everywhere this happens, every publication, every tweet, every post to an on-line forum… they all count as a vote. Do this for whatever planets of the Solar System you think should be included. Ceres, Pluto, Eris, Sedna, Titan, Ganymede, Callisto, etc. If the lexicographers do their jobs properly, this will become the de-facto definition.

What about the "official" definition? There's no such thing, as we've seen with literally every other branch of science. The IAU can create its own classification scheme, but this is not a definition, and it's not official to anyone outside the IAU. Any organization can create their own classification schemes. We have two color schemes — one additive, and one subtractive. We can have two views of planets too, one based on orbits, and another one based on intrinsic properties. There is no law, nor practical reason to forbid this.

But what about the science?

Science requires a healthy degree of skepticism. Every claim must be challenged, and it must withstand those challenges intact. The IAU's definition has clearly failed this process. It is not a scientifically valid definition.

If the science is important to you, then we need to start over and find a system of taxonomy that works for everyone. Keep in mind that more than one kind of science uses these words, and the common language they, and we, all share belongs to no one exclusively.

The science of linguistics is a study of how the masses communicate thoughts, opinions and beliefs with each other. Language is a dynamic means of idea propagation. Those ideas, in the form of words and how we use them, shape who we are as a culture. They shape what is important to us, and to our children. The words we use matter.

There is also a science of educational methodologies. Kids love challenges. How many kids have learned all the state capitals or memorized 40 digits of pi just to demonstrate how smart they are? How much of a challenge is eight planets? It's boring! And it simply doesn't make sense.

Let's give people a bigger, more interesting Solar System to capture our imaginations. *Participate in the debate* of whether a planet should be an object, or a kind of orbit. Given the facts, it's not hard to calculate which side will win that debate.

Or, we can leave it to the science of marketing to grab kids' attention with a thousand Pokemon characters. You decide which is more important.

A Brief List of Sources and References:

People, Blogs, Interviews:

Brown and Stern:

https://www.cbsnews.com/news/demoted-from-planethood-pluto-makes-its-case/

Metzger and Runyon:

https://room.eu.com/news/iau-wrong-about-plutos-planetary-status-say-scientists

Alan Stern:

https://www.forbes.com/sites/jamiecartereurope/2021/02/15/yes-pluto-is-a-planet-says-nasa-scientist-at-the-site-of-its-discovery-91-years-ago-this-week/?sh=204b657068df

https://www.nbcnews.com/id/wbna40470791

https://www.space.com/9553-pluto-planet-experts-weigh.html

https://www.columbusunderground.com/could-pluto-become-a-planet-once-again-jb1/

Lisa R. Messeri

https://www.jstor.org/stable.25677402

C.P. McKay

http://dx.doi.org/10.1016/S0032-0633(00)00051-9

Jean-Luc Margot

https://arxiv.org/abs/1507.06300

Kirby Runyon

https://phys.org/news/2017-02-geophysical-planet-definition.html

Seth Shostak and Iain McDonald

https://www.thedailybeast.com/the-surprising-search-for-the-milky-way s-lost-children?ref=home

Julio A. Fernandez

http://articles.adsabs.harvard.edu/cgi-bin/nph-iarticle_query?1980MNR AS.192..481F&data_type=PDF_HIGH&whole_paper=YES&type=PRI NTER&filetype=.pdf

Alain Maury

http://www.spaceobs.com/Blog-de-Alain-Maury/How-he-believed-he-k illed-Pluto-alone-and-why-it-had-it-coming

150 scientists associated with the New Horizons mission:

https://science.sciencemag.org/content/350/6258/aad1815

https://dx.doi.org/10.1126/science.aad1815

Papers, Publications, Articles:

The Moon is a Planet:

https://www.scirp.org/pdf/IJAA_2017121516131006.pdf

https://www.scirp.org/journal/PaperInformation.aspx?PaperID=81133

https://www.wired.com/2012/12/does-the-moon-orbit-the-sun-or-the-ea rth/

https://www.quora.com/Isaac-Asimov-concluded-that-the-Earths-moon -is-not-a-satellite-of-the-Earth-but-a-sister-planet-Is-this-view-factually -accurate-even-if-culturally-hard-to-accept?share=1

Planet-sized "dwarf planets" in the Kuiper Belt

https://www.sciencemag.org/news/2016/01/astronomers-say-neptune-sized-planet-lurks-beyond-pluto

https://www.space.com/37295-possible-planet-10.html

A super-Earth in the Kuiper zone:

https://www.universetoday.com/152059/astronomers-find-a-huge-planet-orbiting-its-star-at-6000-times-the-earth-sun-distance/

Jupiter does not orbit the Sun

https://www.sciencealert.com/jupiter-is-so-freaking-massive-it-doesn-t-actually-orbit-the-sun

Expert geophysical planetary definition:

https://www.hou.usra.edu/meetings/lpsc2017/pdf/1448.pdf

USGS:

Rivers and streams:

https://www.usgs.gov/special-topic/water-science-school/science/rivers-streams-and-creeks?qt-science_center_objects=0#qt-science_center_objects

Mountains, hills, lakes, and ponds:

https://www.usgs.gov/faqs/what-difference-between-mountain-hill-and-peak-lake-and-pond-or-river-and-creek?qt-news_science_products=0#qt-news_science_products

IAU:

Resolutions:

https://www.iau.org/administration/resolutions/general_assemblies/
https://www.iau.org/static/resolutions/Resolution_GA26-5-6.pdf

Guidelines for naming astronomical objects:

https://www.iau.org/public/themes/naming/
https://minorplanetcenter.net/iau/info/DesDoc.html

Recommended Books

Beyond Pluto,
John Davies ISBN 0-521800196

The Case For Pluto,
Alan Boyle ISBN 978-0470505441

Impact Clan,
Dr. Germano D'Abramo ISBN 978-1518735332

Chasing New Horizons,
Alan Stern, David Grinspoon, et al. ISBN 978-1250098962

The Pluto System After New Horizons,
S. Alan Stern, Jeffery M. Moore, et al. ISBN 978-0816540945

To:

Family,
909199 Our Street
My Town,
City, State,
USA, Earth,
Zonecode
411-237-404-888-016

PostCard

From PLUTO

Greeting from Pluto! You can't see it in this photo, but our hotel is at the base of the tallest mountain in the picture. The skiing is amazing and the view of charon is truly lovely. It's huge! And it's always in the same place in the sky, it's impossible to get lost. That's a big relief to your Father. No need to ask for directions! You wouldn't believe the views of the milky way from the observation domes. This is the best place in the solar system to get away from it all. Having a wonderful time, wish you were here.

Love, Mom.